I
Petrified Carbon

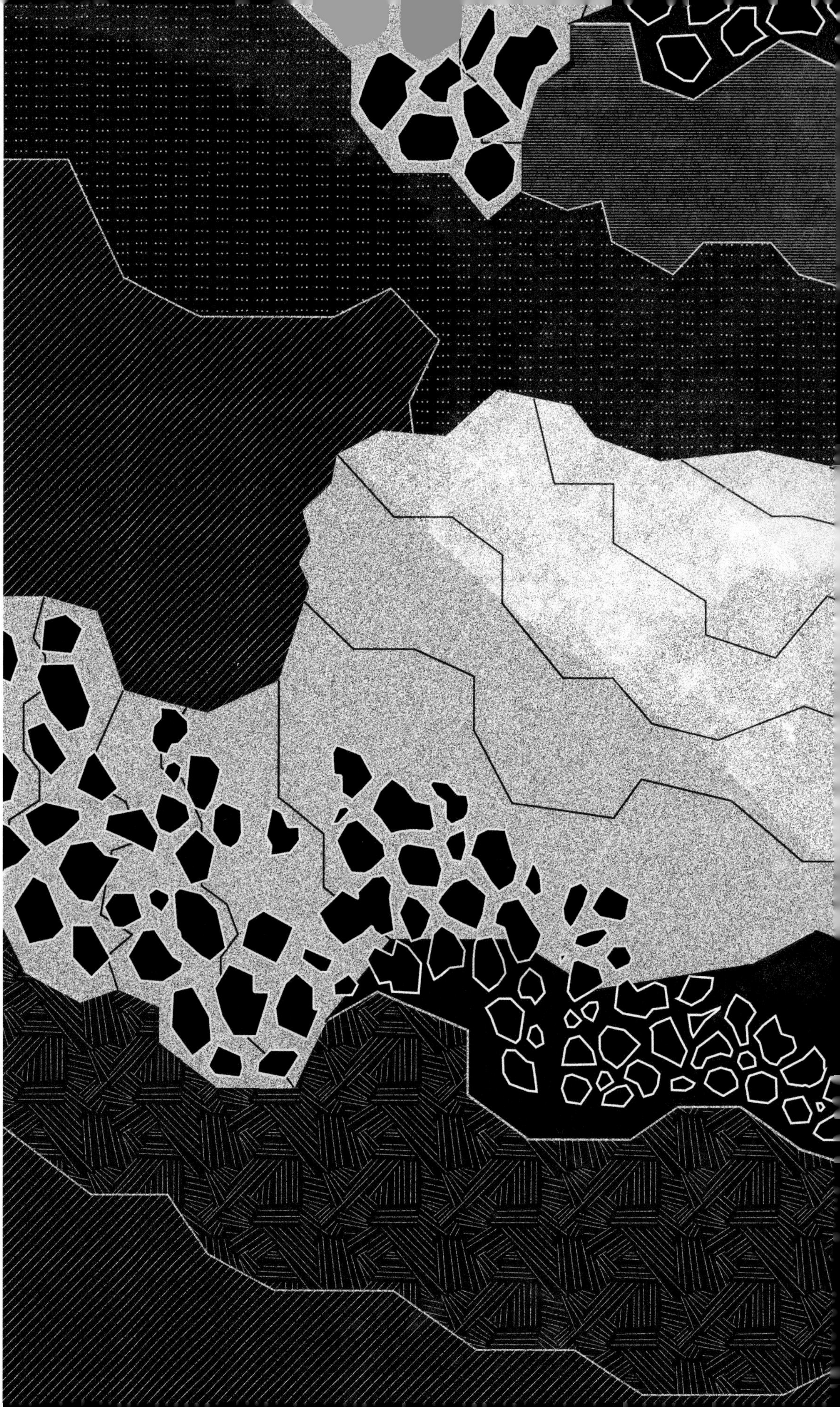

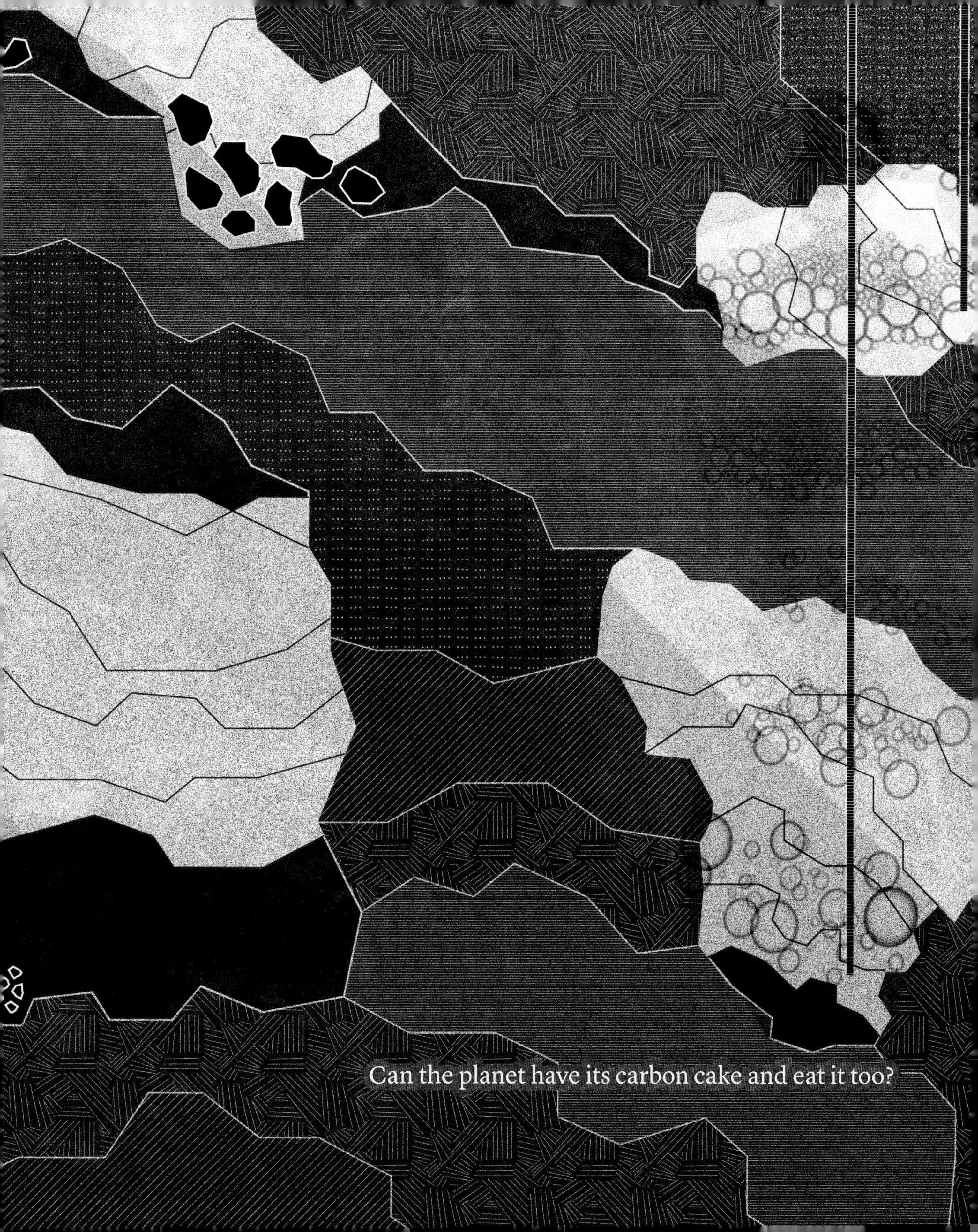
Can the planet have its carbon cake and eat it too?

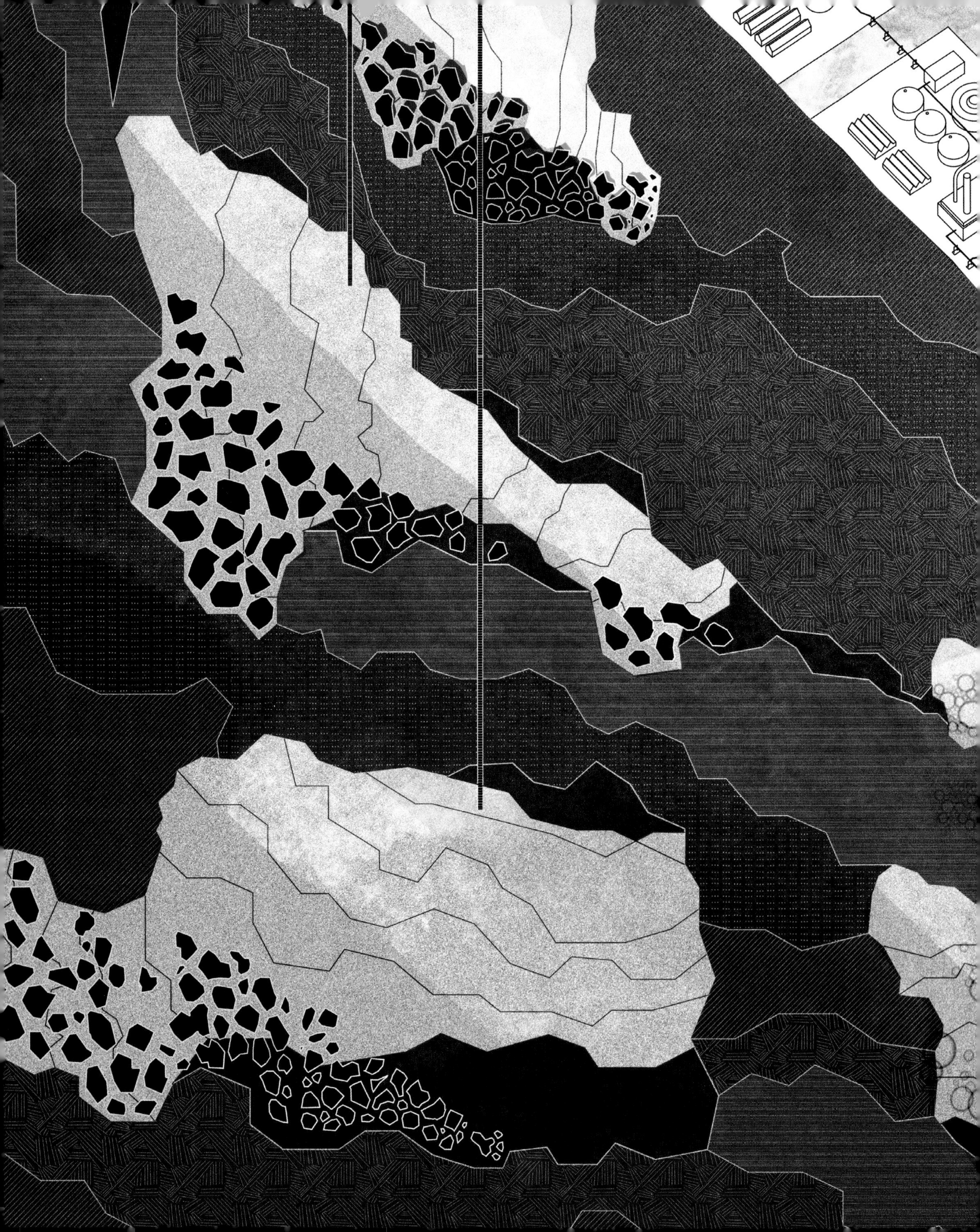

The fossil fuel industry promotes CO_2 capture and sequestration in deep geological formations.

In fact, oil companies pioneered this process of injecting CO_2 into porous subsurface rock, known as enhanced oil recovery, in an effort to squeeze more oil that cannot be extracted otherwise; and have commercially developed the technology through five decades of pollution and funding of climate change denialism.

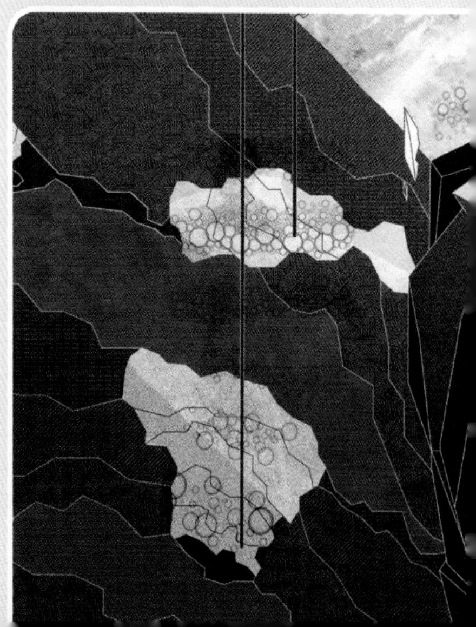

Depleted oil and gas reservoirs are early sites of carbon sequestration projects.

More oil out;
more carbon dioxide in;
more revenue for the largest polluters.

It’s all upside!

The Earth trembles and cracks under pressure.

CO2 Leakage from
Geological formed faults

The oil industry repurposes its pipelines and tankers into a carbon storage infrastructure.

Natural Climate Mitigation is the new Beyond Petroleum.

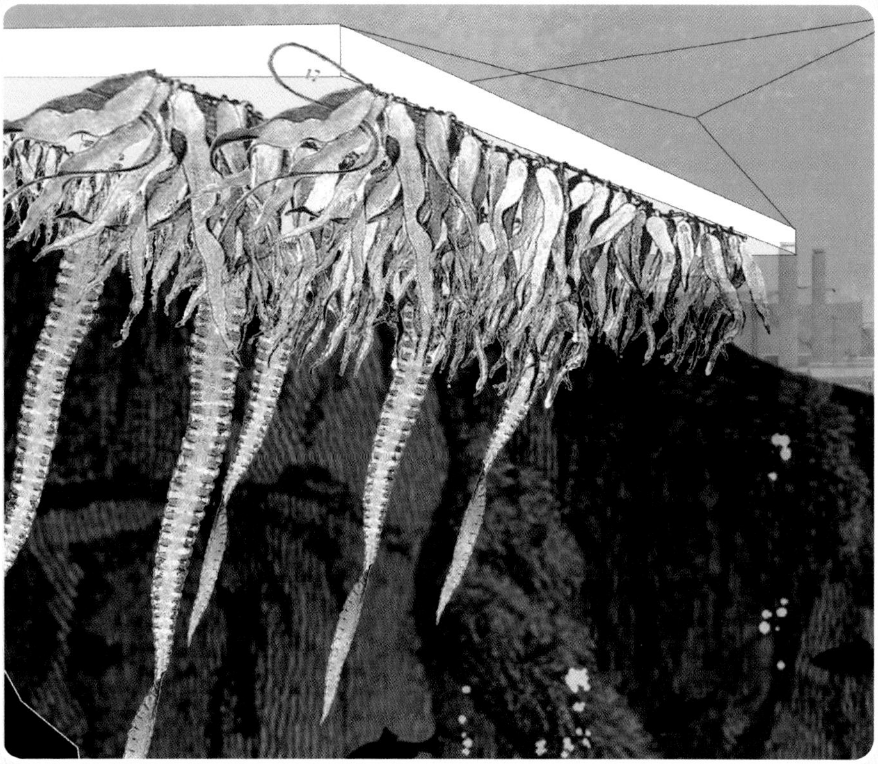

Fully loaded tankers roam the oceans capturing carbon dioxide emissions in open water algae plantations.

The oceans, Earth's largest carbon sink, absorb a third of humanity's greenhouse gas emissions. The water becomes more acidic and harmful to marine life.

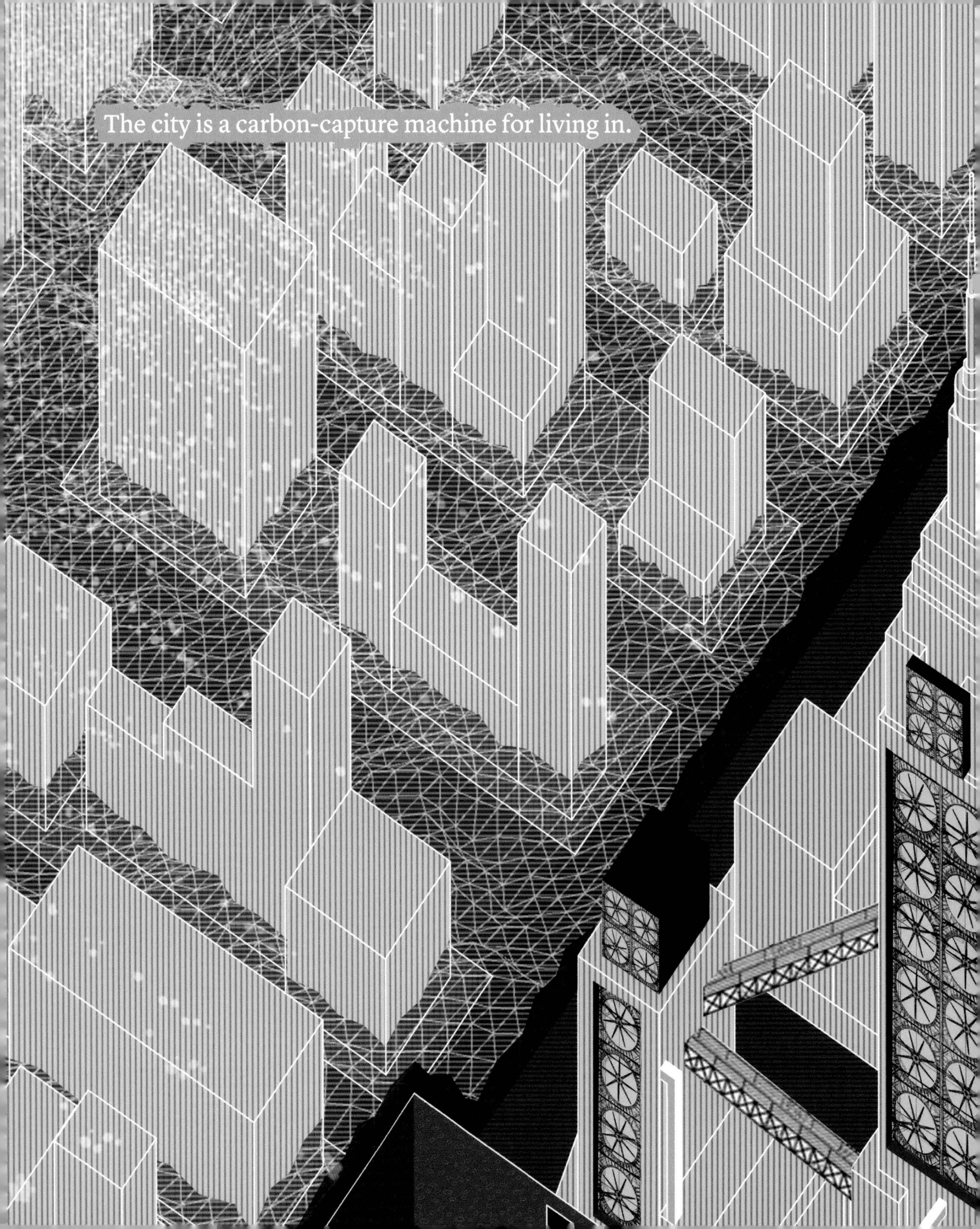
The city is a carbon-capture machine for living in.

RCP 6.0 - Some Mitigation
RCP 2.6 - Aggressive Mitigation

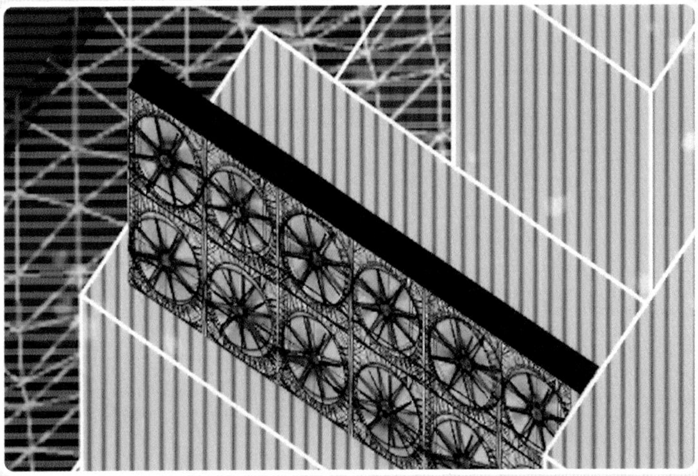

Smog-eating towers capture carbon dioxide. A fleet of carbon scrubbers runs down facades.

Clean air is released to the atmosphere. All other substances are stored for reuse in the construction industry.

CO2 synthetic cement
CO2 carbon synthetic diamond

Industrial vacuum plants sky-mine carbon dioxide and convert it into baking soda, bleach, and other products.

A large volume of CO_2 is pulled beneath, to the City of the Captive Carbon Globe.

A repurposed Cold War nuclear shelter, the City is a recovery plant that compresses carbon in a high-pressure, high-temperature process and turns it into diamond.

The carat is the new measurement unit of the carbon footprint.

The Carbon Clock is the Sky with Diamonds. The higher the atmospheric CO_2 level, the brighter the stars. The diamond crystals have a yellowish color, probably because of nitrogen impurities in the process.

City air makes one (carbon) free, they say.

II
Arctic Albedo

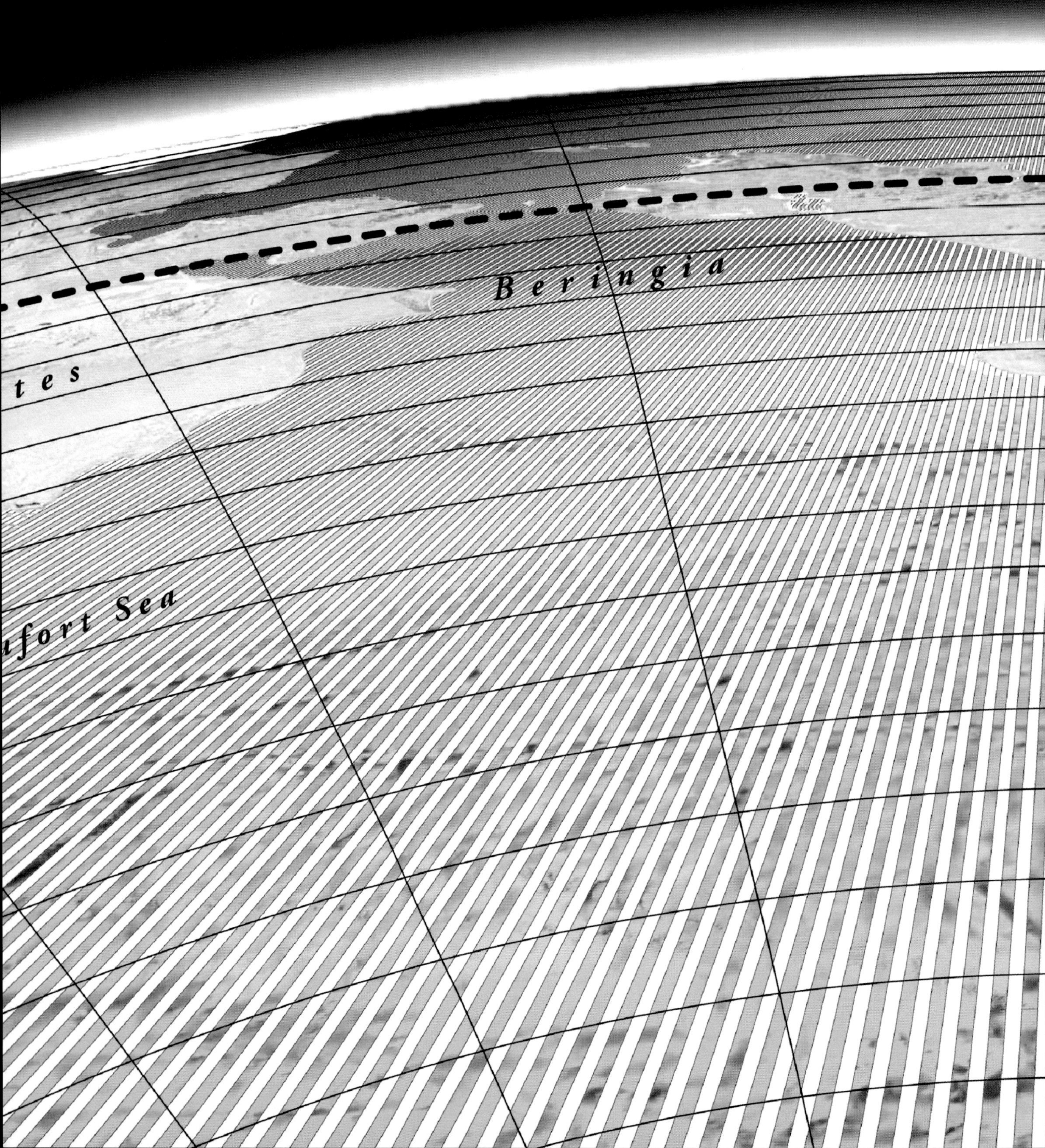

Beringia
tes
ufort Sea

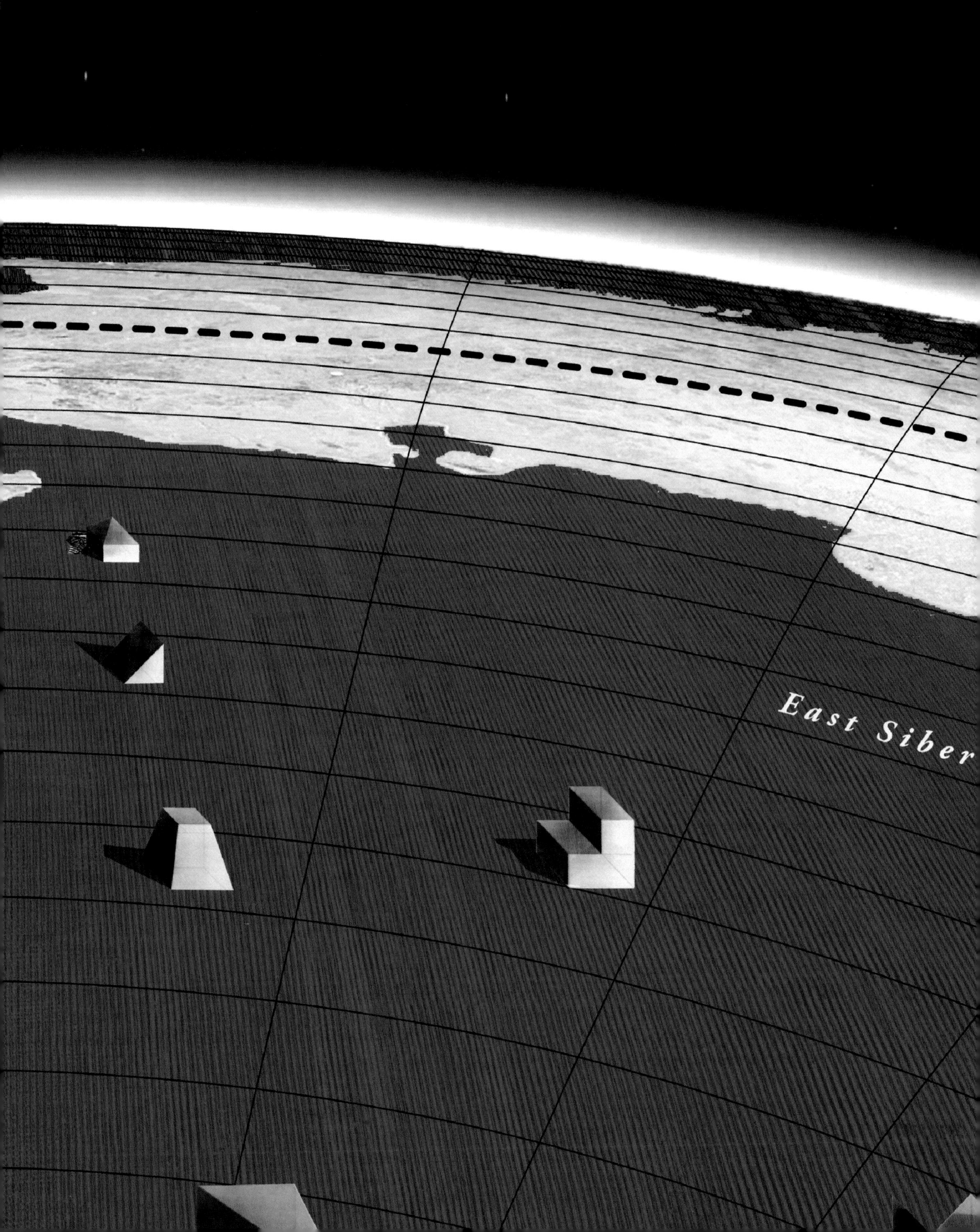
East Siber

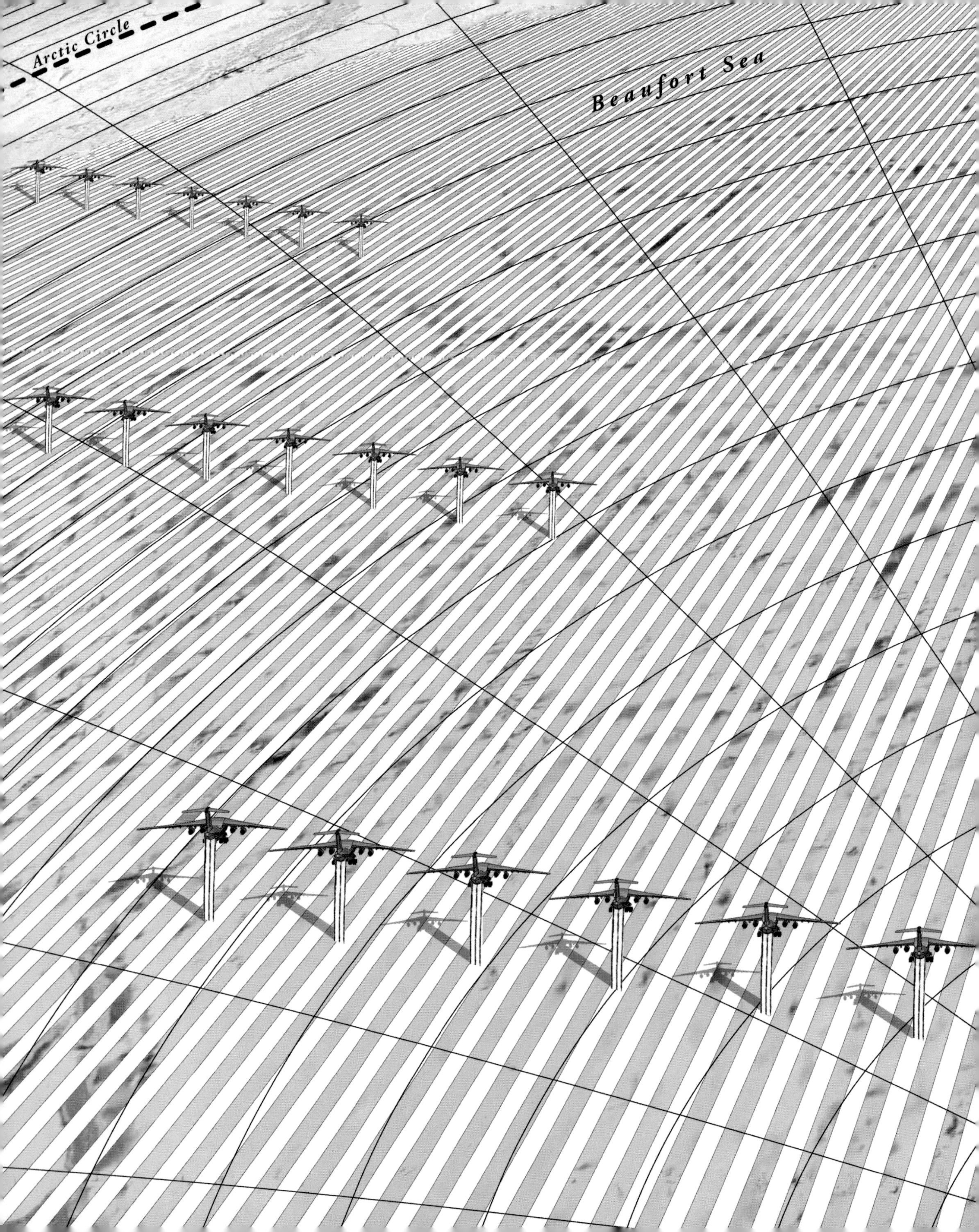
Arctic Circle
Beaufort Sea

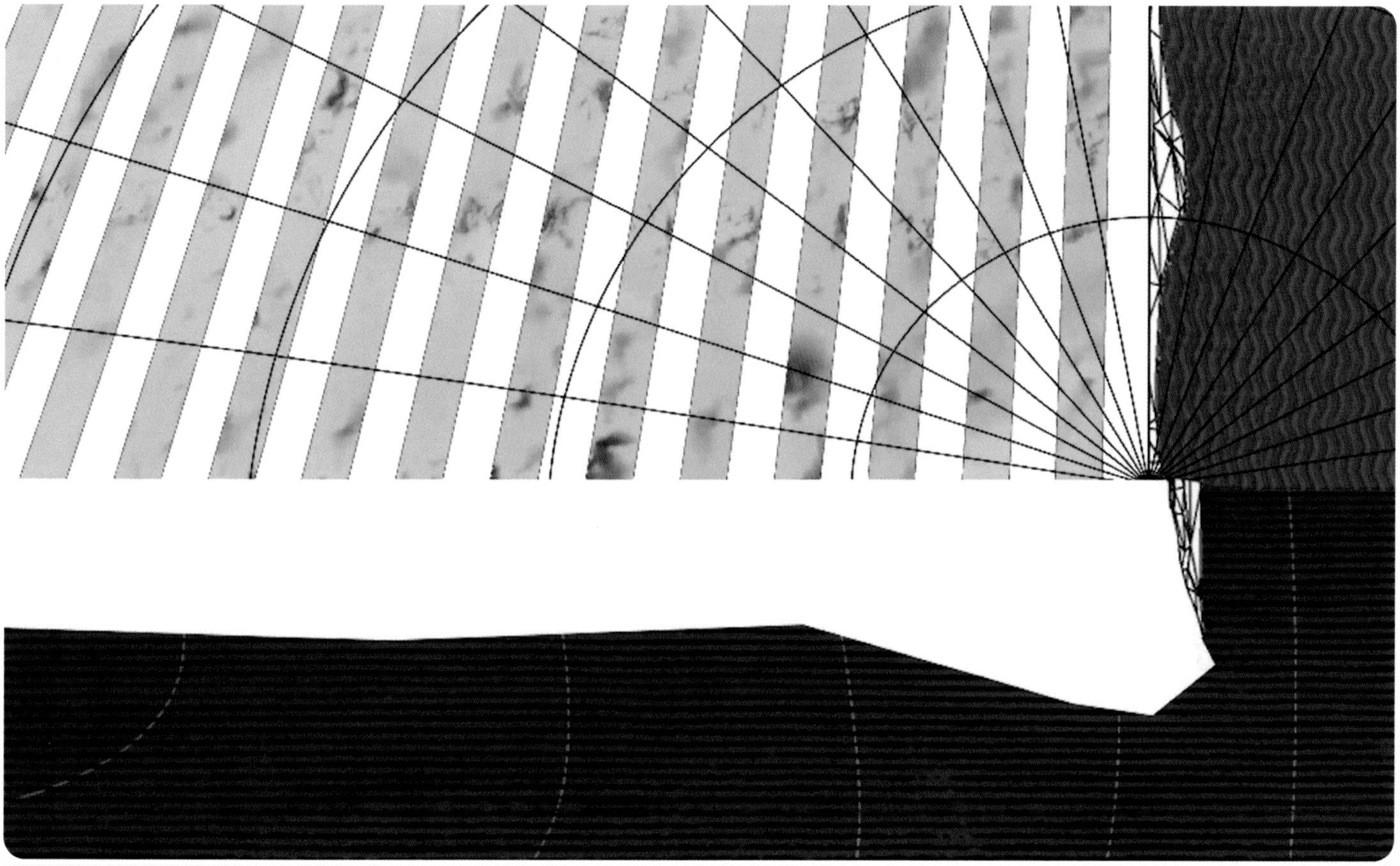

The Arctic is warming twice as fast as the rest of the planet.

The melting of the snow and ice cover reduces the Arctic's albedo effect. The exposed darker areas of seawater and soil absorb more solar radiation leading to further warming.

Bands of small glass beads are applied to slow the meltdown.

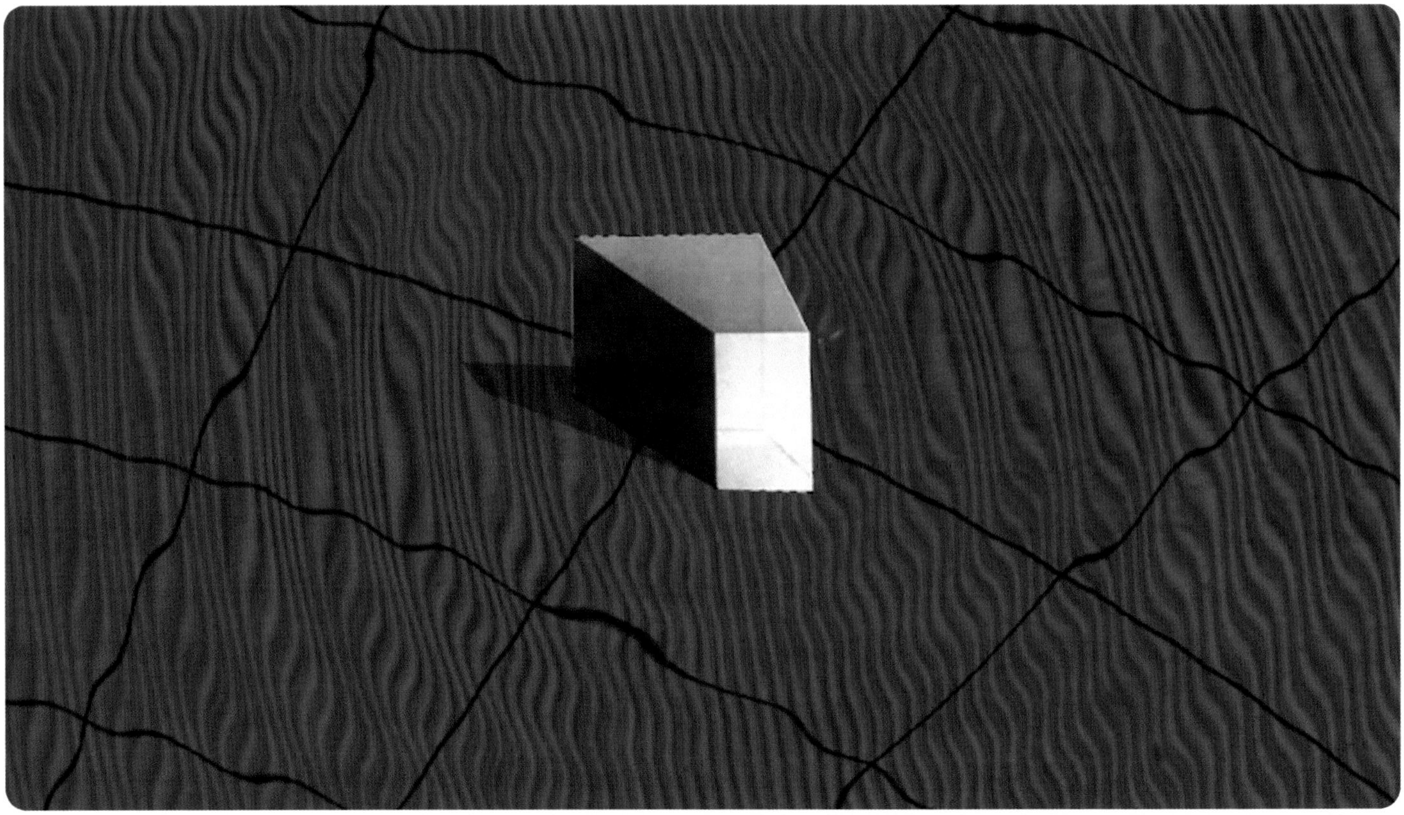

A blue ocean event—an ice-free Arctic summer—looms on the horizon.

Less sea ice is more ship traffic.

Vessels from Arctic Council members plant their flags on the seabed to lay claims to oil and gas reserves below the ice in an auction sale fit for Jules Verne's Topsy-Turvy, *The Purchase of the North Pole*.

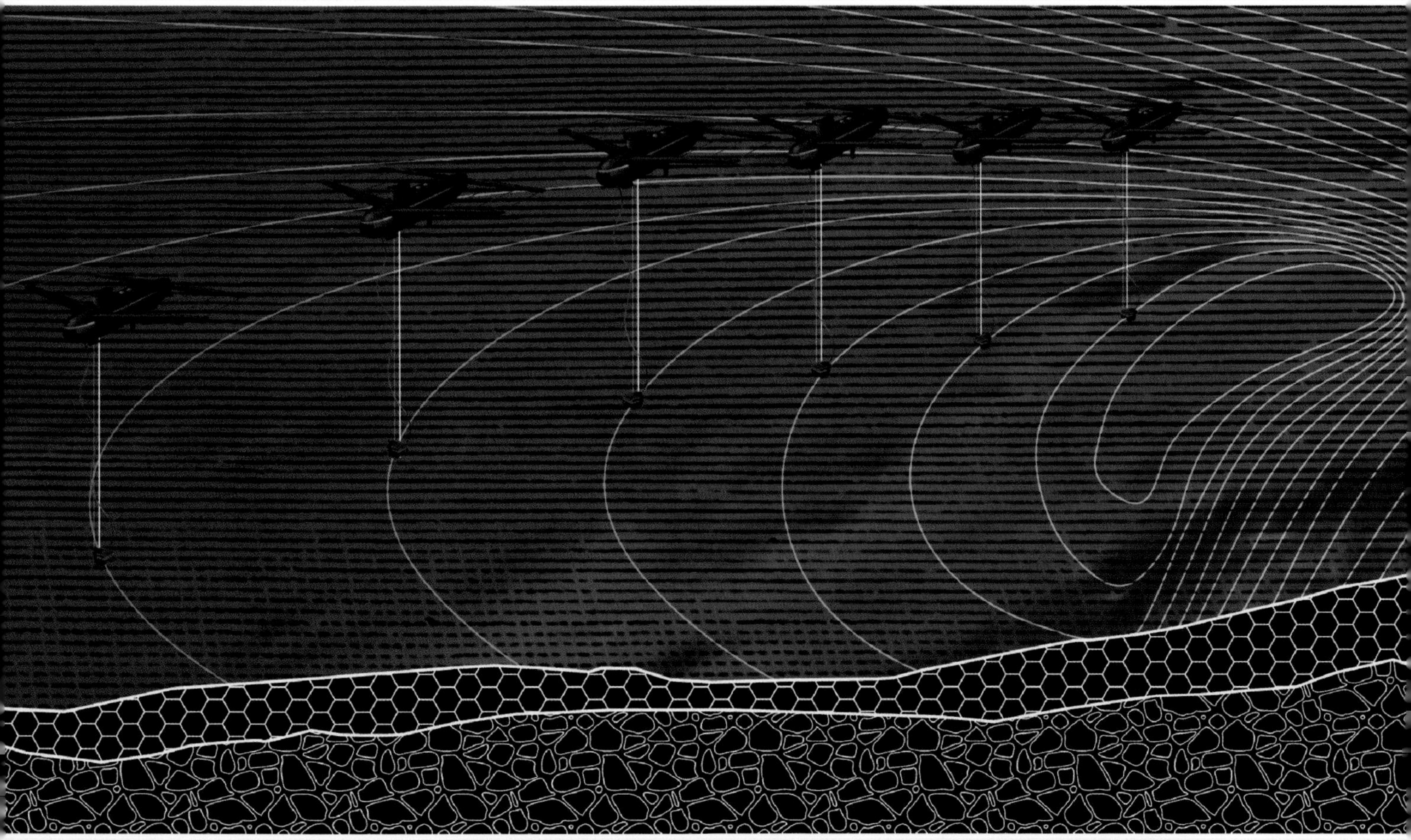

Permafrost thaw releases hundreds of millions of years of buried carbon into the atmosphere.

To restore the Earth's white pole, seawater is pumped onto floating hemispherical structures creating artificial glaciers.

Methane that bubbles up through soil and water is trapped in such biomes. Methane-eating bacteria convert the greenhouse gas into biotechnological applications.

WOOLLY R HINOCEROS 13500
STEPPE BISON 13000
MEGALOCEROS 12000
SAIGA 11500
OTH 10500

Rewilding the Arctic slows the climate crisis. Herds of large herbivores—horses and bison—are released onto the tundra to create vast grasslands.

Extinct woolly mammoths, revealed by the melting permafrost, are cloned and introduced as well.

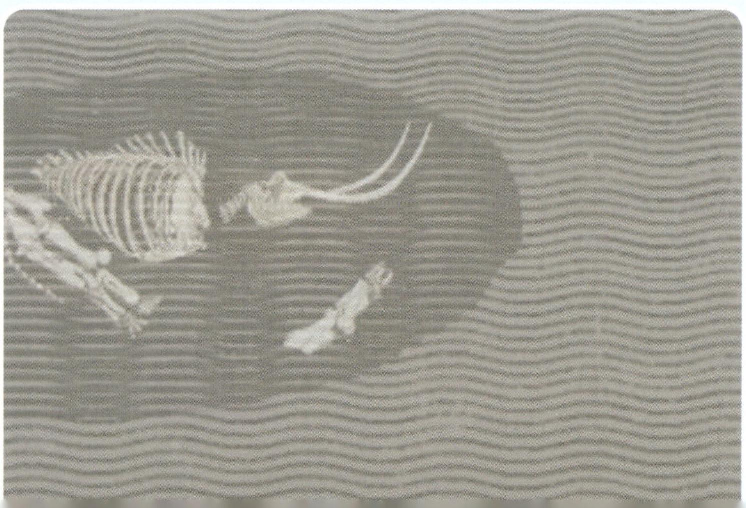

In the ‘mammoth steppe’, the grazers stomp thick topsoil and trample on snow protecting the carbon-rich permafrost.

Extreme wildfires convert the old-growth boreal forests from a carbon sink to a carbon source.
High albedo advocates further clear the forest to boost reflectivity and expand linear monocultural plantations.

The woodland caribou is in the cross-fire; its lichens winter forage is destroyed.

Indigenous Guardians rekindle with prescribed burns setting off controlled fires within cleared perimeters.
The Word for World is (Still) Forest.

Layer
frost

The Alaska-based transmitter facility harvests methane emissions into a giant snow globe. An array of antennas sends high-energy beams to ionize methane into pure carbon and hydrogen.

The energy intensive turbines are powered by natural gas drawn from the North Slope Petroleum Reserve.
Climate change deniers collect even more fortunes from raising the Arctic albedo.

Not another aurora borealis—tiny particles of diamond fall from the sky.

III
Sky River

Rainmaking artificially induces precipitation by dispersing substances into the clouds.

Cloud seeding was weaponized by the U.S. Air Force during the Vietnam War—to soften military supply roads surfaces and cause landslides—and was later banned by the United Nations.

Extreme droughts and global warming have spurred a renewed interest in the pattern modification of vapor flows.

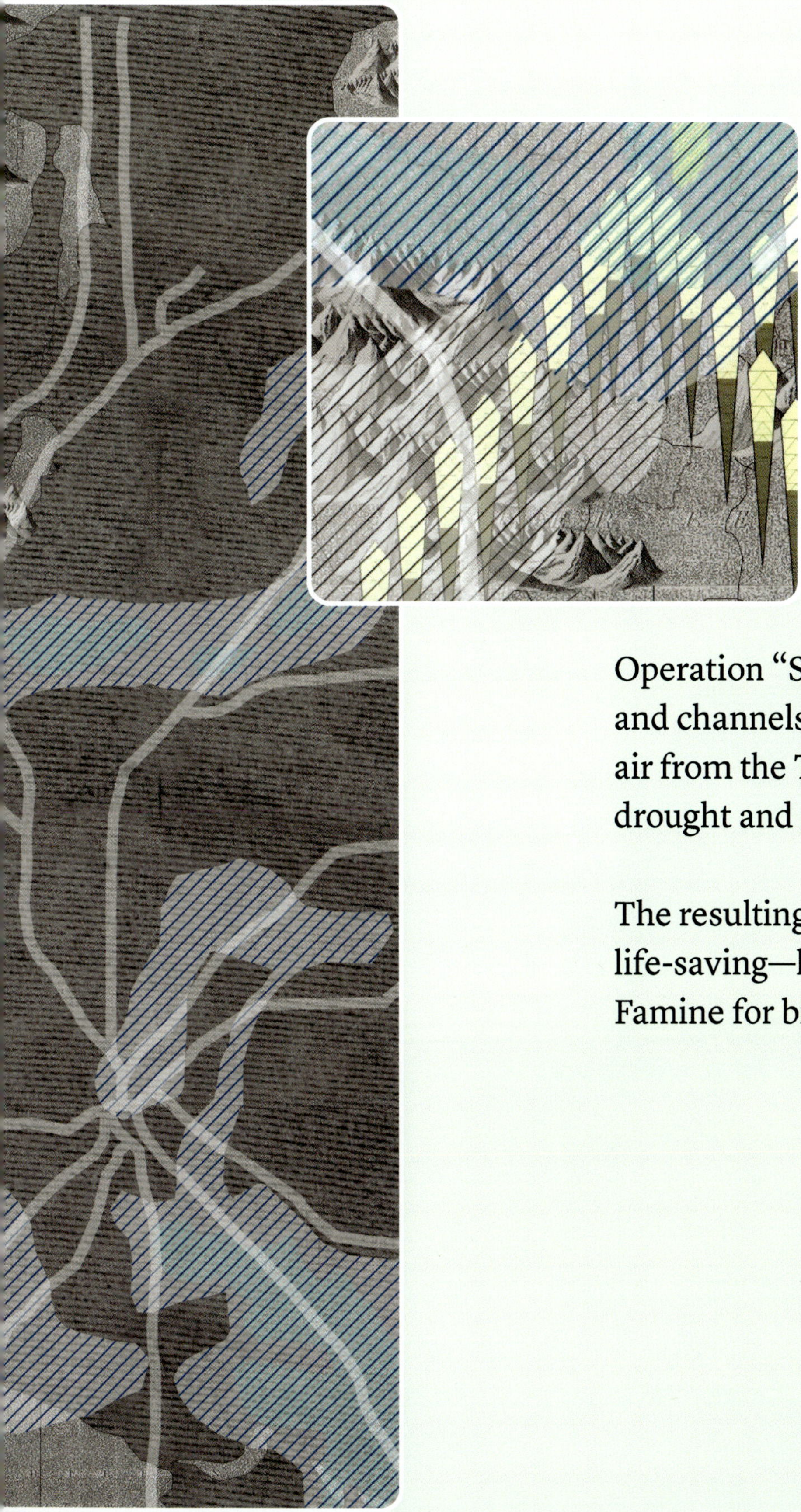

Operation “Sky River” melts snow and channels high-moisture bands of air from the Tibetan Plateau to mitigate drought and irrigate the Gobi Desert.

The resulting rain is promised to be life-saving—literally avoiding a Great Famine for billions of people.

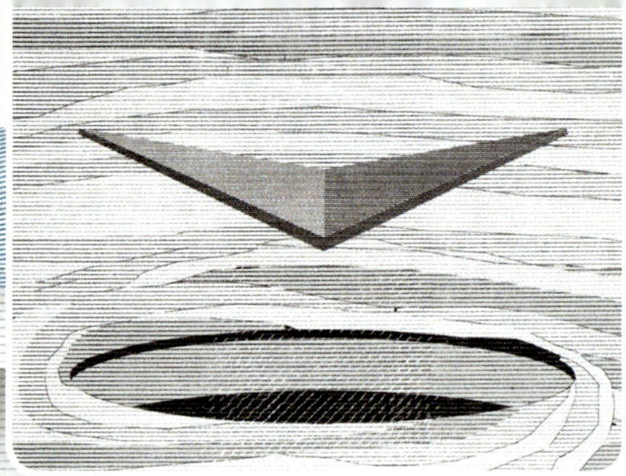

Thousands of crystal-like fuel-burning chambers blast silver iodide into the atmosphere.

Weather satellites provide real-time data to track sky river patterns.

Political strife grows between highlands and lowlands as they compete to secure freshwater resources.

Water wars are fought in international courts.

Storm Kings perform ritual dances and light great fires to invoke rain.

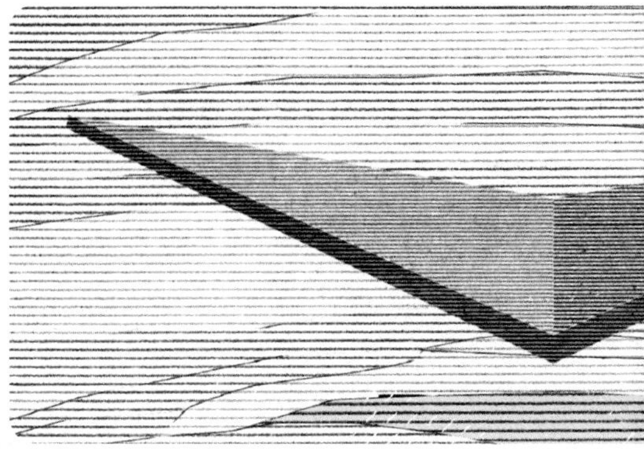

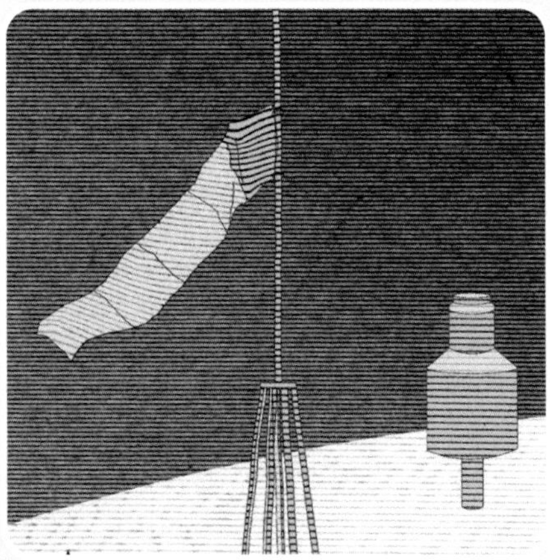

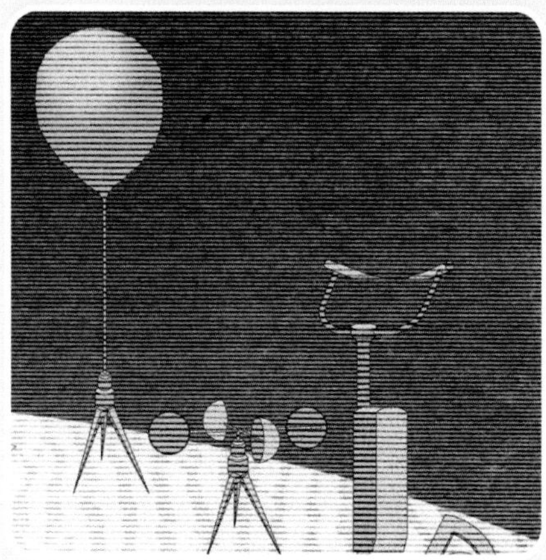

The rainmakers of today prime clouds with particles of silver iodide.

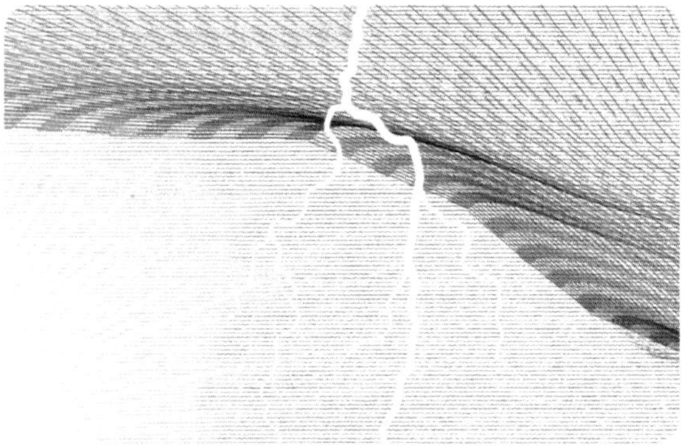

The water cycle is a zero-sum game: increased rainfall in one region is moisture missed elsewhere.

In countries that heavily rely on the river basins, the Kings quarrel the skies for rain.

The free silver ion is extremely toxic in aquatic environments.

A Spiral Jetty of dead juvenile fish is sometimes visible depending on the water level.

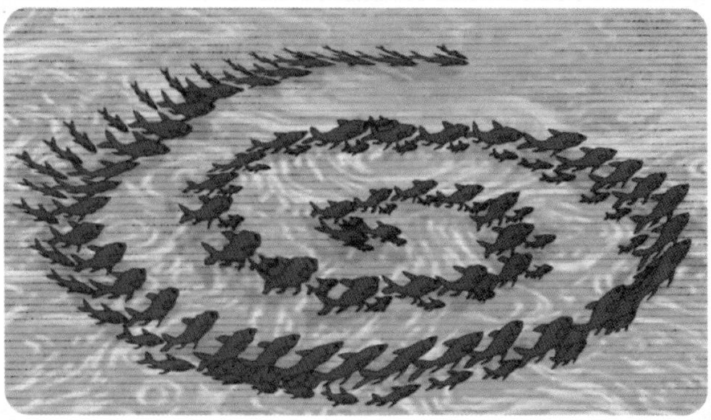

IV
Sulfur Storm

Nuclear Power Russia
Taal, Philippines
Pinatubo, Philippines
Mayon, Philippines

North Pacific Ocean
Equator

Climate scientists debate the use of solar radiation management techniques.

Edward Teller, “father of the hydrogen bomb,” proposes injecting sulfate aerosol into the atmosphere to reflect sunlight away from Earth.

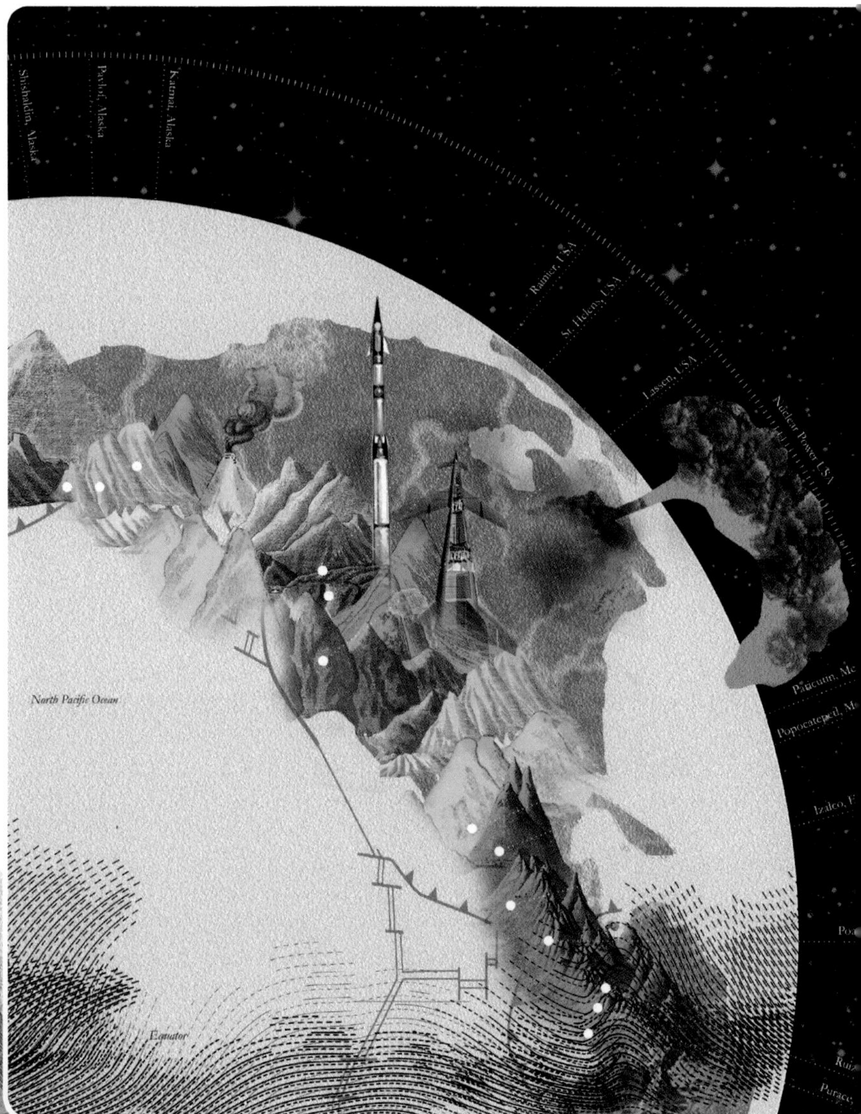

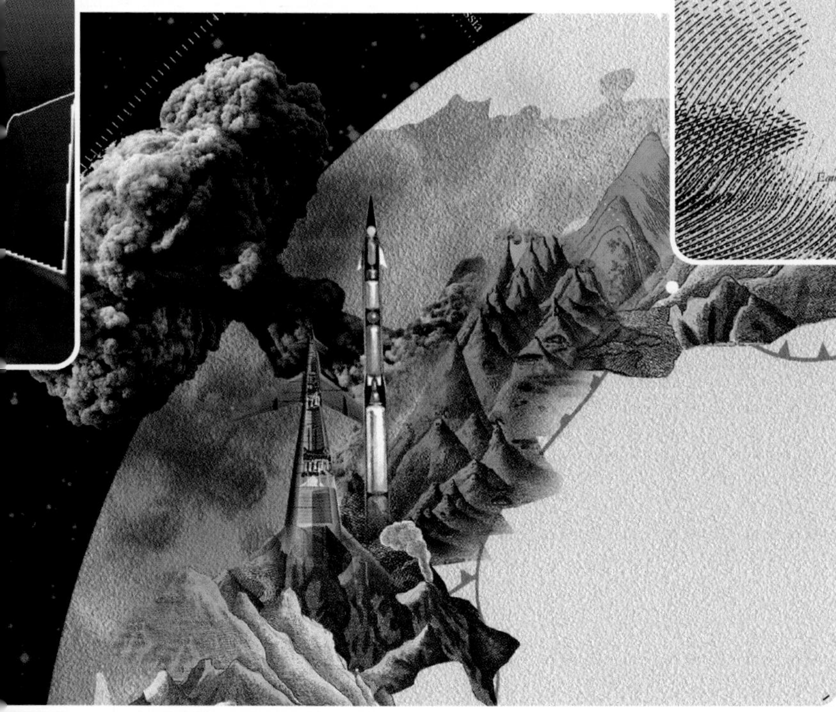

The thickening shroud of particles mimics the solar shield of volcanic dust to recreate the global cooling effects of the 1815 eruption of Mount Tambora.

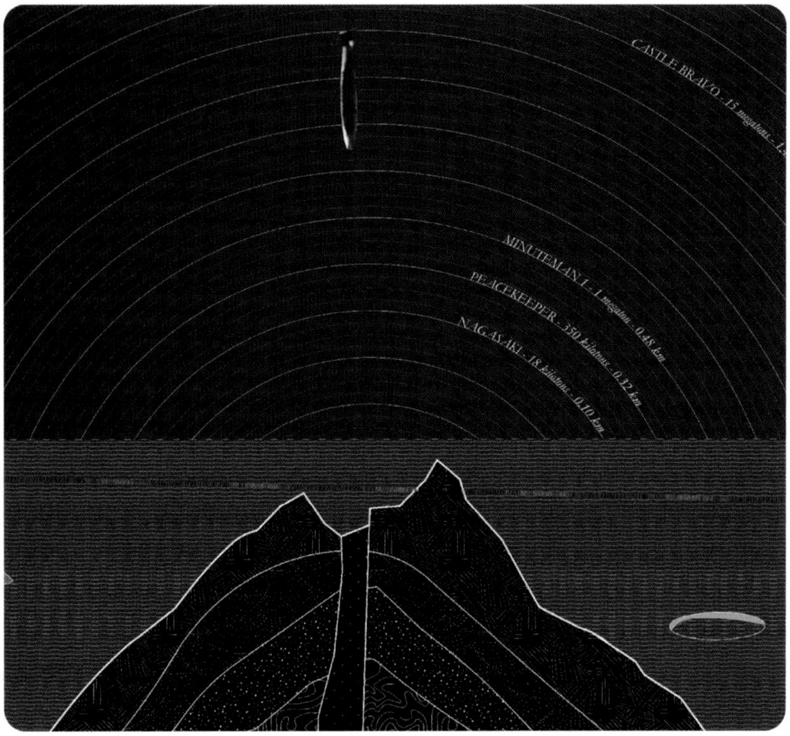

Operation Sulfur Storm detonates the Cold War nuclear arsenal in the magma chambers of tropical volcanoes.

The UN Convention on the Prohibition of Military or Any Other Hostile Use of Environmental Modification Techniques is yet to rule on whether such artificial volcanoes save the climate or destroy the planet.

INDIA 2008
50 launchers
60 warheads
ISRAEL (2002)
70 launchers
270 warheads
PAKISTAN (2009)
70 launchers
90 warheads
FRANCE (2008)
98 launchers
300 warheads
CASTLE BRAVO - 15 megatons - 1.42 km
MINUTEMAN 1 - 1 megaton - 0.48 km
PEACEKEEPER - 350 kilotons - 0.32 km
NAGASAKI - 18 kilotons - 0.10 km
CHINA (2011)
138 launchers
178 warheads
UNITED KINGDOM (2011)
180 launchers
275 warheads

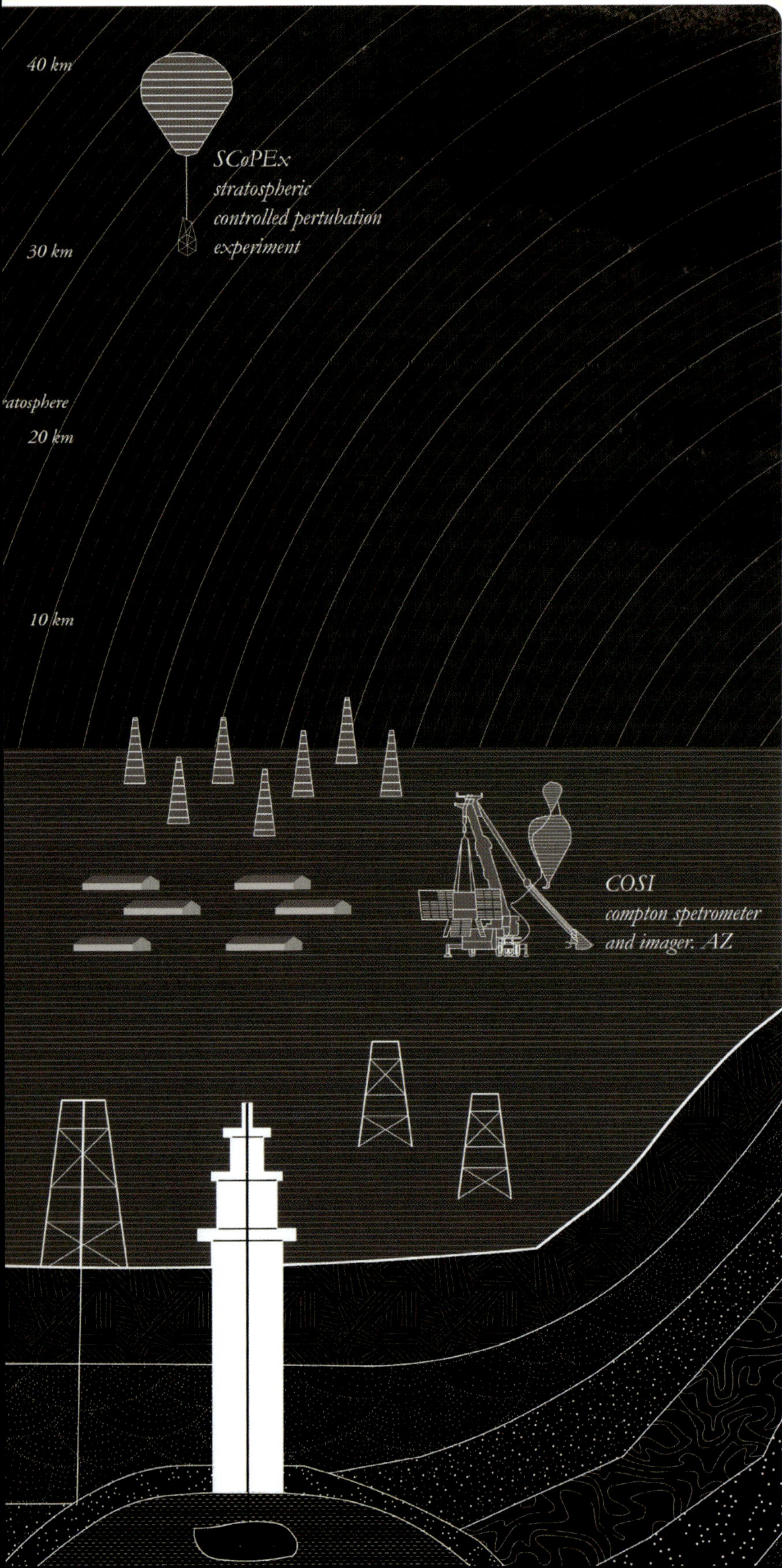
40 km
SCoPEx
stratospheric
controlled pertubation
experiment
30 km
20 km
10 km
COSI
compton spetrometer
and imager. AZ

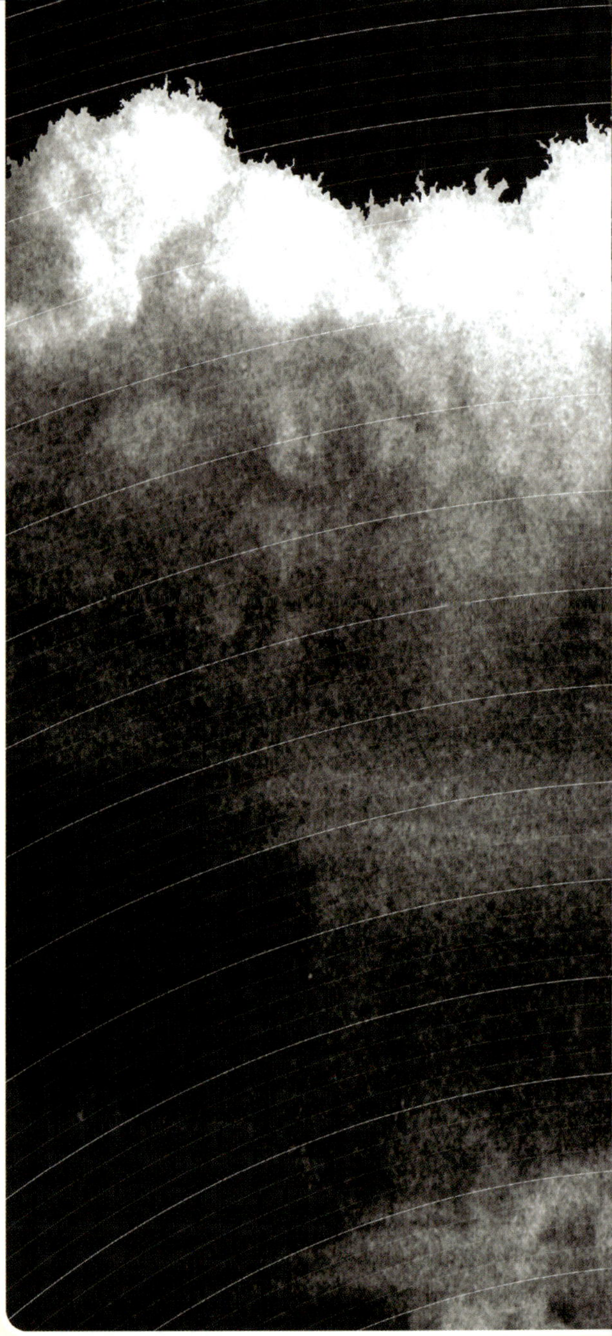

Blue skies become a little less blue.

Balloons, kites, drones, aircrafts and naval artillery continuously spew millions of tons of tiny sulfate particles into the sky. A bright afterglow is a familiar sight at sunset.

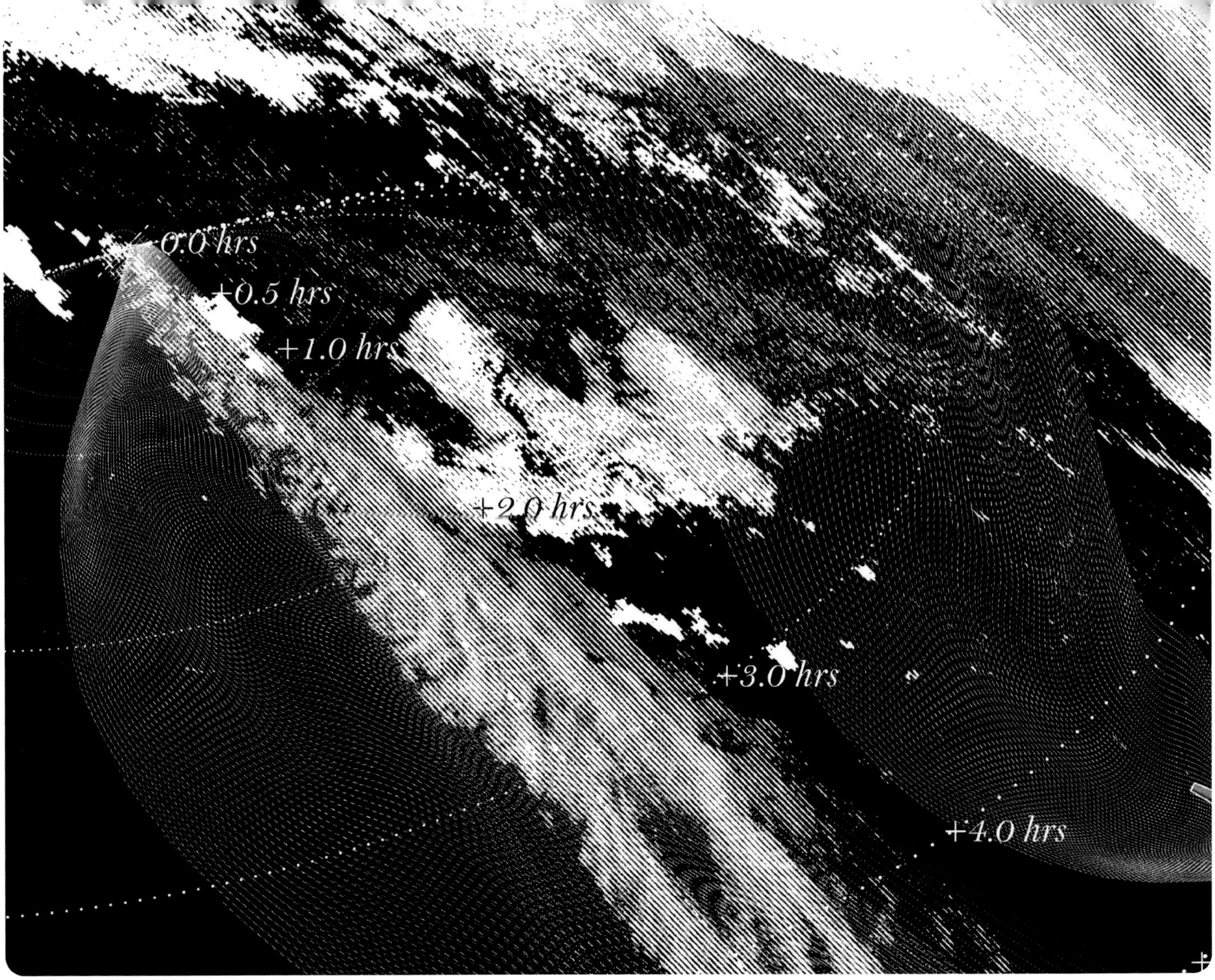

The cloud gravitates into equatorial orbit.
Saturn-like rings circle the Earth.

Clouds of ash deplete the ozone layer reversing the signs of recovery that followed the 1987 Montreal Protocol.

The exhaustion of the ultraviolet shield has dire implications for life on Earth.

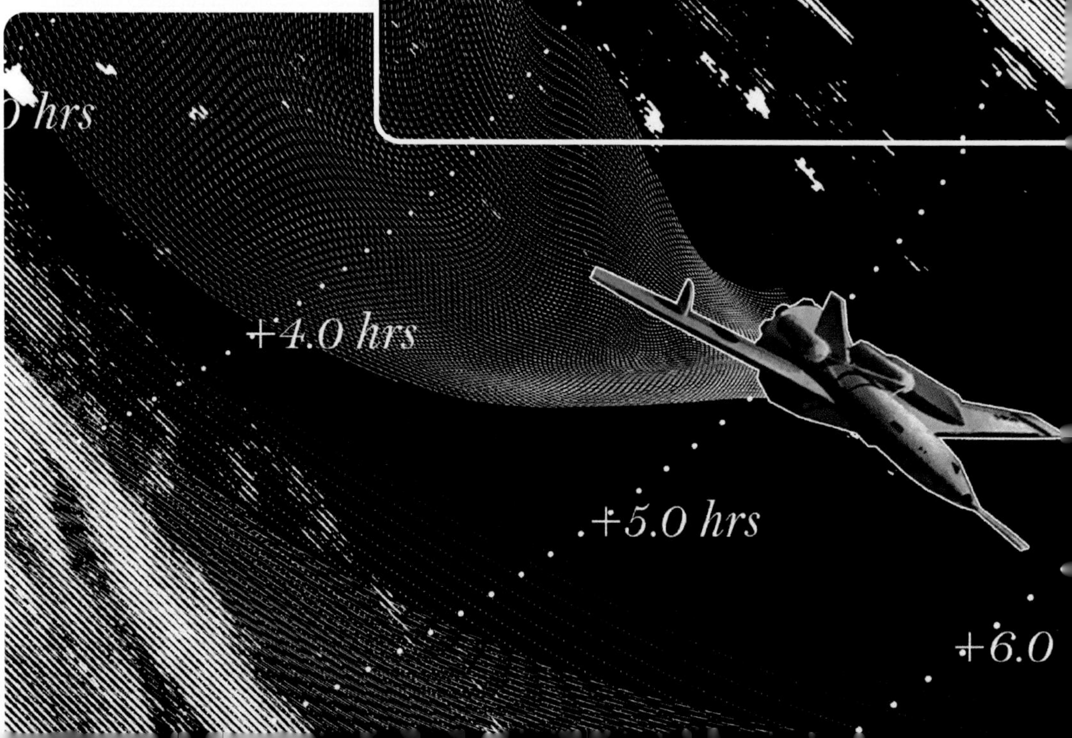

Aerosol particles make it hard to breathe.

A Crystal Vivarium protects from the Sulfur Storm. The sealed container is a refuge for the one and a half-billion winners of geoengineering.

Four times this number are left standing outside.

The Continuous Conveyor Drywall City converts excess sulfur dioxide into synthetic gypsum—wallboard, plaster, and soil additives.

A by-product of coal-fired power plants, gypsum is certified Green Label Plus—local and recycled!

Sulfur dioxide inhibits photosynthesis. The same sulfate production line filters the air for sensitive species.

V
Dust Cloud

A shield that protects the atmosphere from two percent of solar radiation offsets the warming effects of greenhouse gas emissions.

A giant space shade orbits in equilibrium at a Lagrange Point between Earth and the Sun.

The shield is an artificial constellation of dust that is sourced from near-Earth asteroid mines.

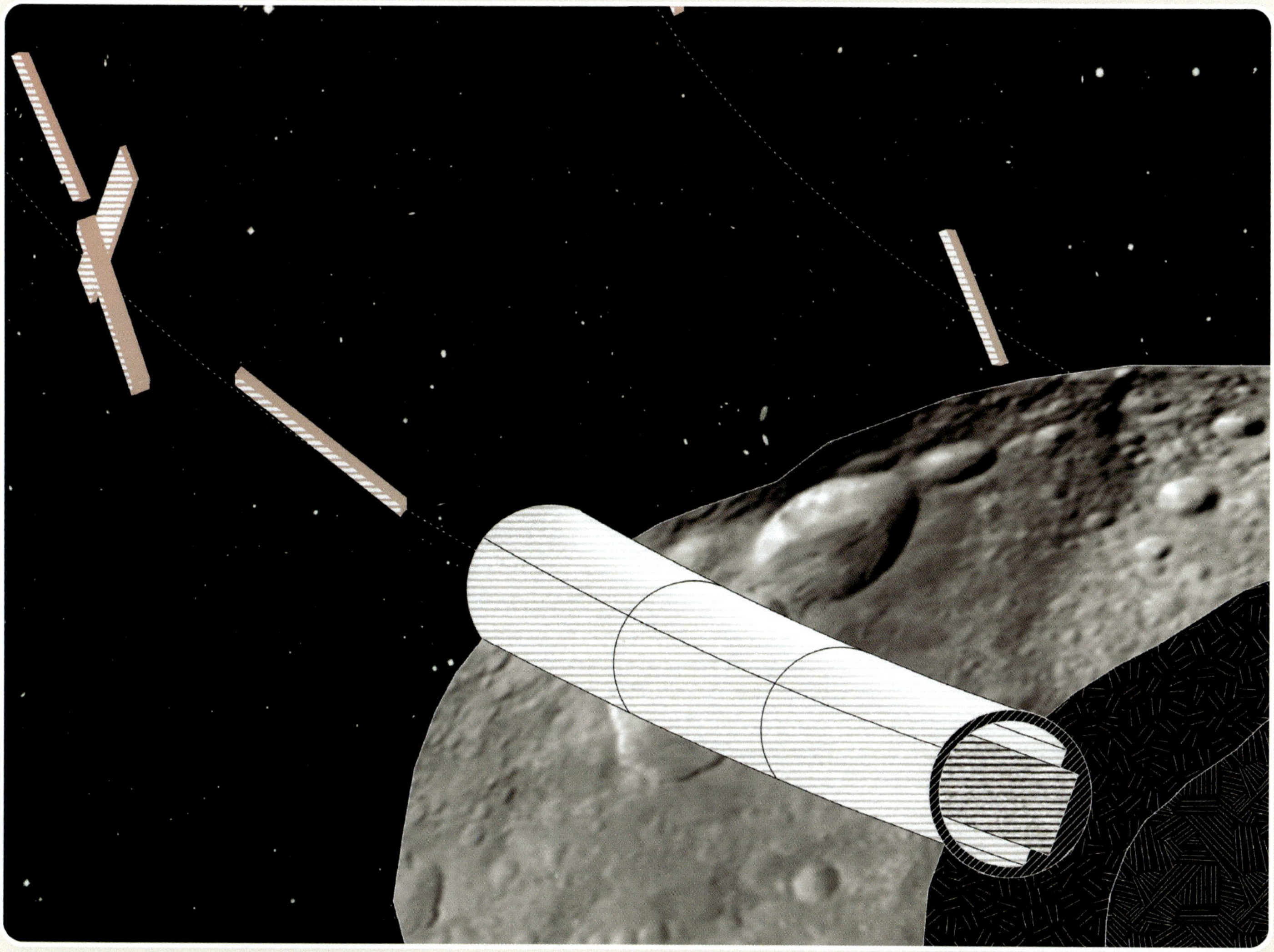

Do you remember the falling stars that traced through the heavens?

A swarm of capsules filled with asteroid dust is launched into the Lagrange point.

The empty capsules cross paths with Earth's orbiting junkyard of satellites. Large robotic arms extend to hit the capsules, as if with a bat, into another million-mile dust trail.

The dust cloud started as an experiment to cool Venus's climate for a future where humans seek refuge on other planets.

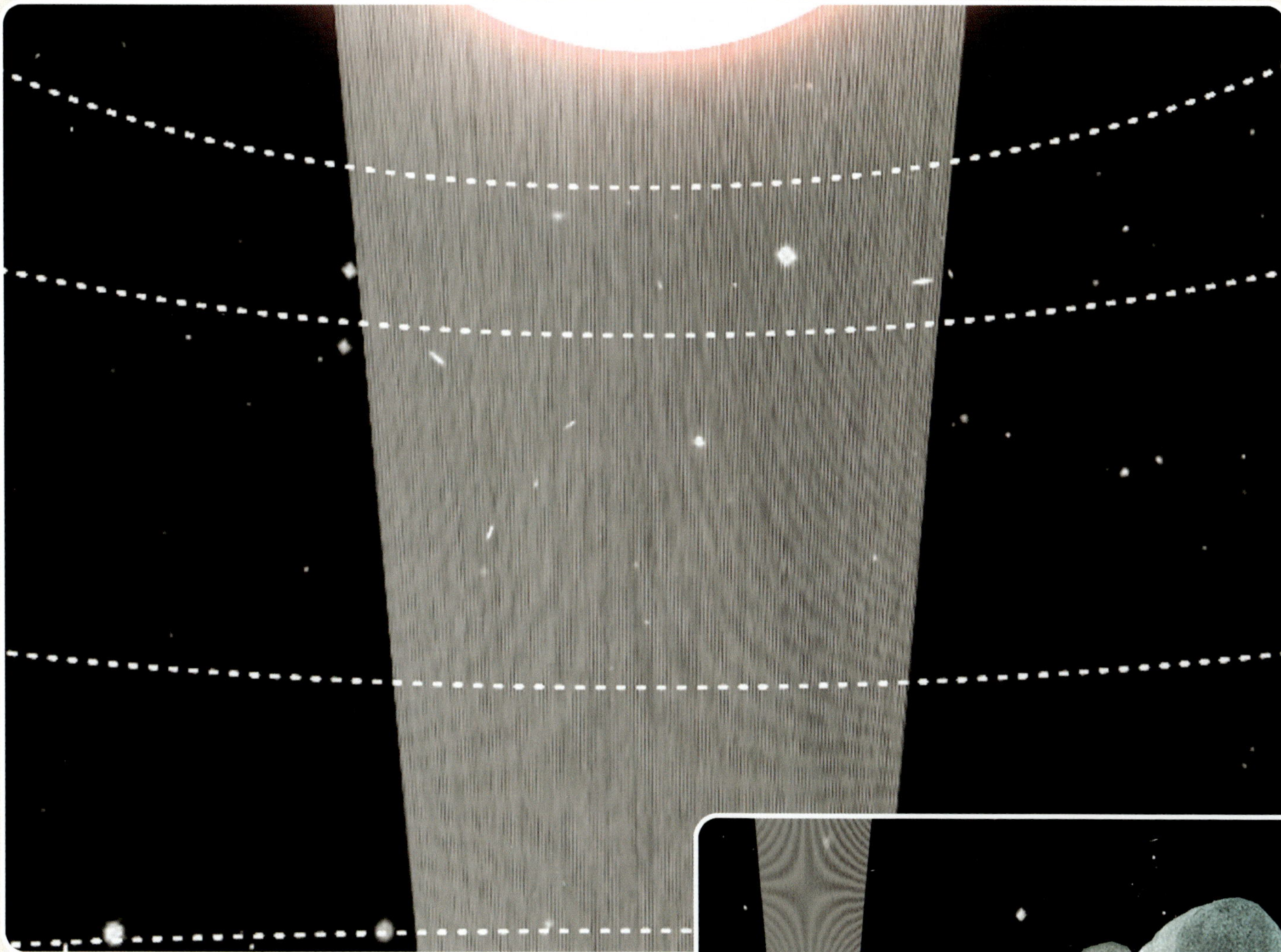

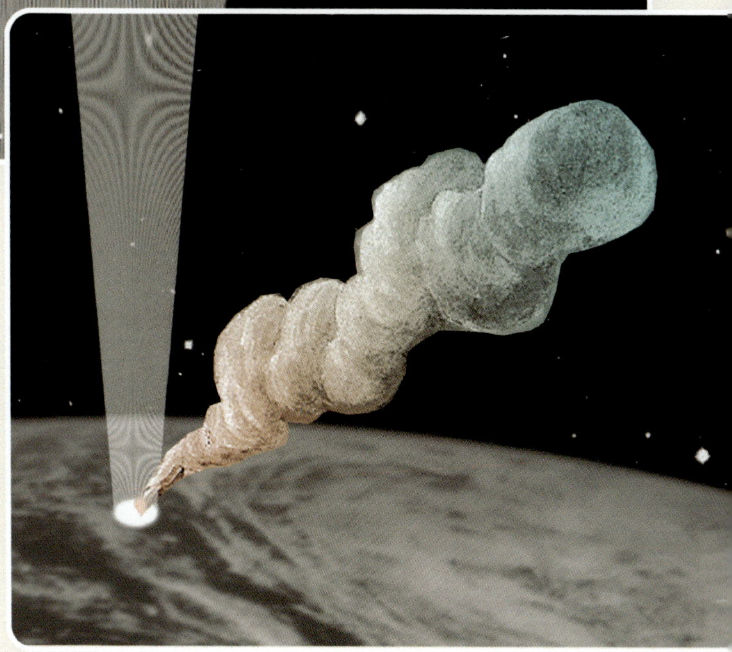

The compacted junkyard was later polished into a space-based solar power plant that beams energy to Earth.

The heliobeam is also a sun gun that concentrates enough heat onto a point to make an ocean boil or a city burn.

The accumulating dust and smoke dull the starry sky.

EUROPA
VOLGA
Bern
ALPS
4 807
DANUBE
5 633
4 160
Rome
Ankara
Athens
Tunis
MOUNTAINS
TIGRIS
ELBURZ
Téhéran
Tripoli
Jérusalem
790
Bagdad
ARABIA
S A H A R A
Rhyad
TIBESTI
3 406
Khartoum
LAKE
CHAD
N'Djamena
AFRICA
MT. CAMEROON
Douala
CONGO
RIFT VALLEY
Kinshasa
EASTERN
RIFT

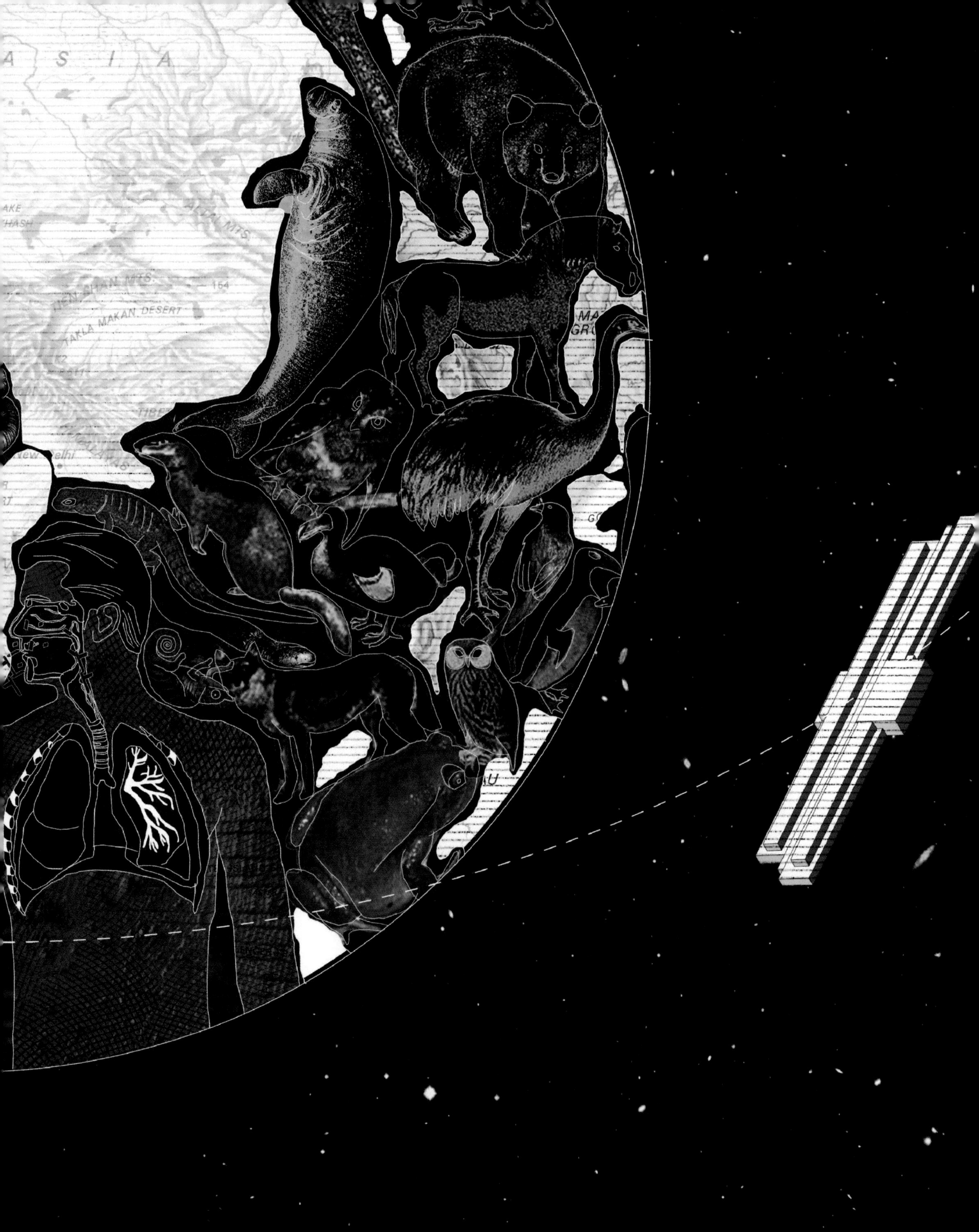
A S I A
TAKLA MAKAN DESERT

I can't breathe.
All wars on life begin by attacking the
respiratory tract.

Living species choke with the carbon damage
that humans wreak on the Earth's lungs.

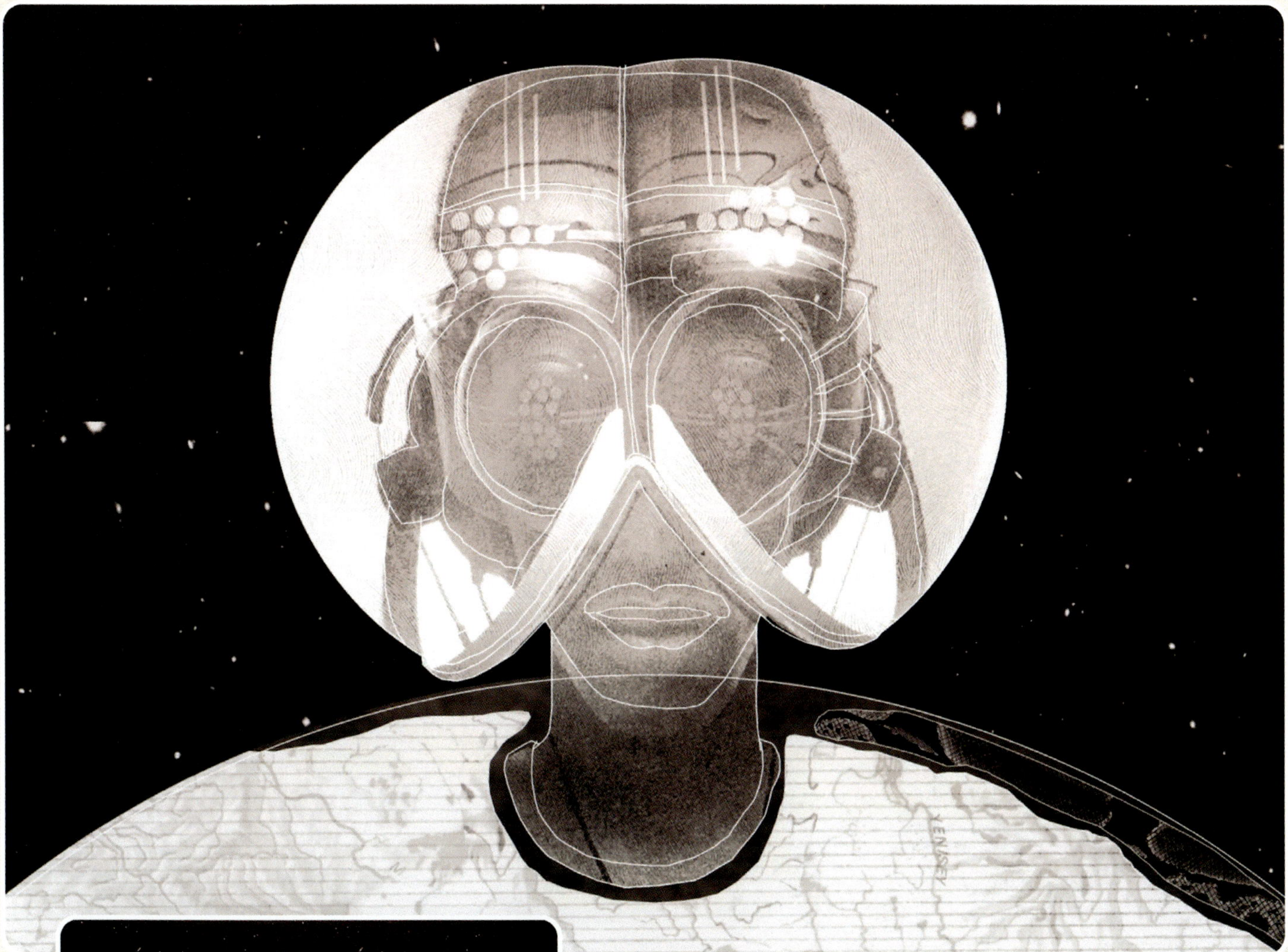

The Respirator Goddess blows the breath of life into the planet.

The terror of geoengineering is to be forced to live intimately with the death of the atmosphere contemplating suffocation as a real possibility.

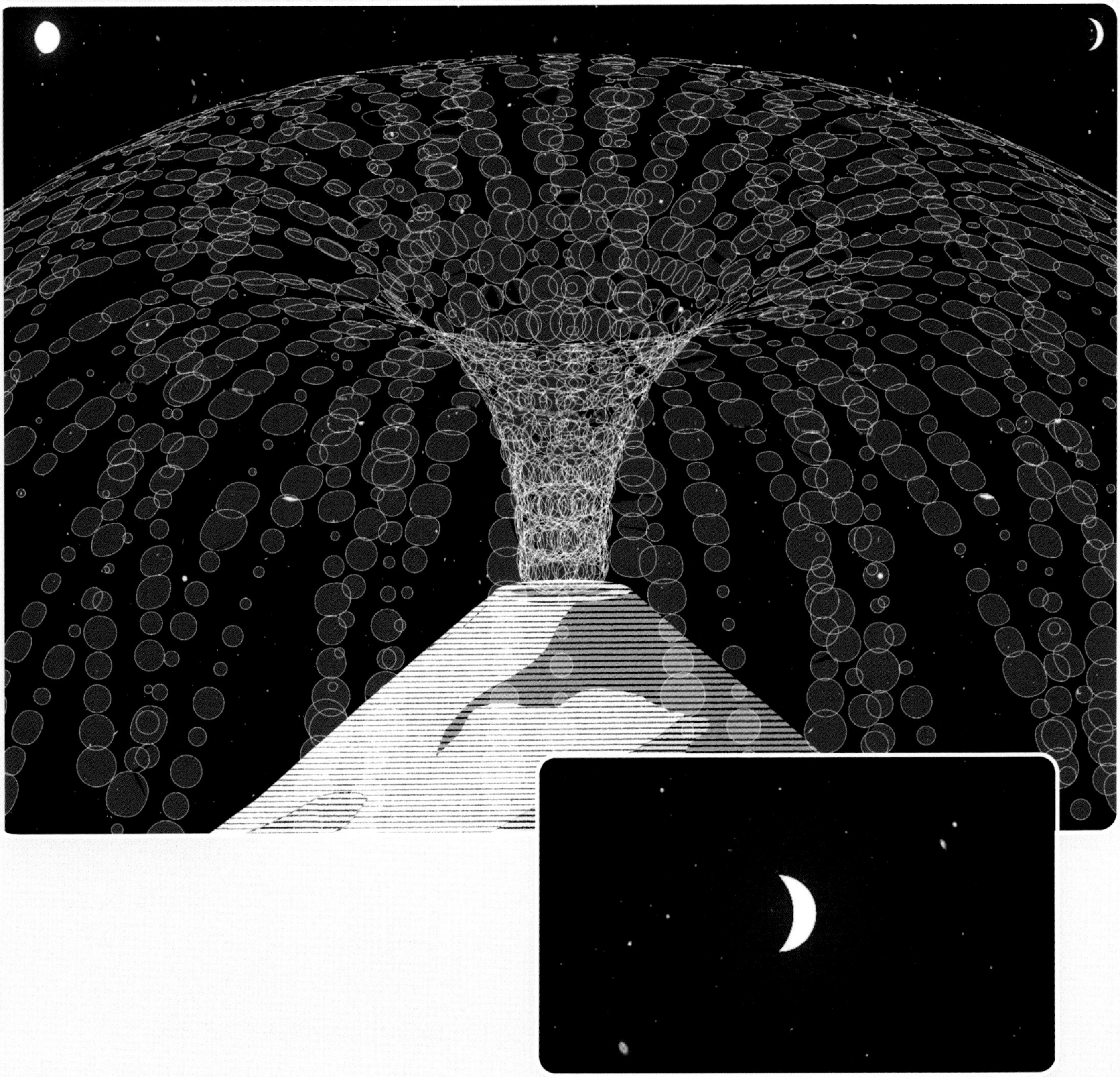

Dust plumes close in an artificial eclipse.

Much of humanity is caught in the stranglehold of sulfur and dust storms.

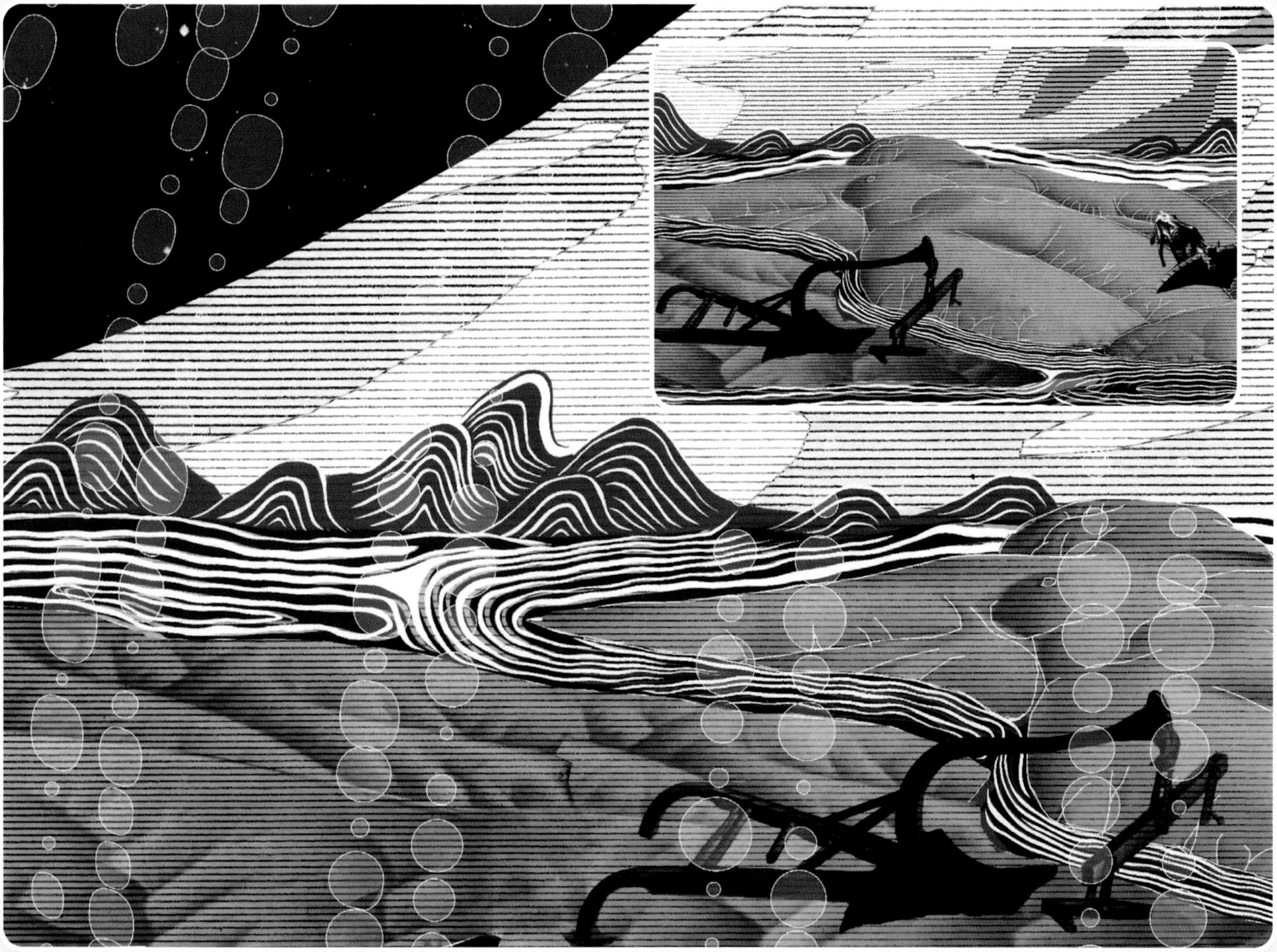

Swirling whorls of dust sweep through barren lands.
Dustomonia; it howls and groans, it squeals and moans.
Every blow produces more pain.

The Earth deserves palliative care.

Welcome to the climate emergency!

The term *geoengineering* refers to technologies that counteract the effects of anthropogenic climate change by deliberately intervening in Earth systems. The climate science community remains divided on whether the planet should be after geoengineering at all. In the midst of a climate crisis, and amidst increasingly widening positions on the future of the warming planet and what to do about it, *The Planet After Geoengineering* employs a speculative fiction approach to think with and against geoengineering as a form of planetary management. The graphic novel makes climate engineering and its controversies visible; it brings the abstract worlds of climate models to public debate to interrogate humanity's responsibility towards Earth and each other.

In five geostories, *The Planet After Geoengineering* assembles a planetary section that cuts through the underground, crust, atmosphere, and outer space. Each geostory—Petrified Carbon, Arctic Albedo, Sky River, Sulfur Storm, and Dust Cloud—depicts possible future Earths that we come to inhabit on the heels of a geoengineering intervention all while situating such promissory visions within a genealogy of climate-control projects from nineteenth-century rainmaking machines and volcanic eruptions to Cold War military plans.

Such fabrications of an engineerable earth open a space to forge a new geopolitics that includes the actual Earth—its dimensions, processes, and lifeforms—as constitutive concerns of planetary design. The graphic novel grapples with some issuing questions: which lifeforms are in need of being safeguarded, from what threat, by which actors and means, to which ends, and at which costs. The aspiration of this work is that it can galvanize a shift towards public communication that explicates the climate crisis as it invites responses to the climate emergency.

The Geoengine

Kathryn Yusoff

Kathryn Yusoff is a professor of inhuman geography at Queen Mary University of London.
She is author of *A Billion Black Anthropocenes or None* (University of Minnesota Press, 2018).

As a companion to a Design Earth's speculations on the politics and practices of geoengineering, this essay outlines a framework for thinking about the 'geoengine' that underpins concepts of planetary modification. It reviews some of the ways in which geoengineering has been framed as a technological, governance, and promissory discourse and examines the contradictions inherent within some of these framings. Questioning some of the motivations and forms of participation that are evident in the governance of geoengineering, and its current challenge to existing forms of global democracy, the essay speculates on the wider geopolitical implications of operating at the scale of the planet, and what this means for how we currently understand the 'geo' in geopolitics as a material-discursive-transformative practice. This essay suggests that only by attending to the geo-ontological formations inherent in the fabrication of an engineerable earth, can a new geo-politic be accommodated within political thought that takes account of a much more active 'geoengine' that is being opened to political and material modification. And, more broadly, what kinds of technics are at stake in the anticipatory geologics of making new geophysical worlds?

Working with the exploratory quality of geoengineering to produce 'new climates,' Design Earth's "The Planet After Geoengineering" employs a speculative approach to thinking with and against geoengineering as a form of planetary management. What this project foregrounds are the visualization of the underlying motivations and imaginaries of geoengineering and a speculative mode of apprehending and interacting with inhuman earth processes. Cutting across the project's five chapters is an investigation into the forms of 'geo' circulating in the knowledge economies and practices of geoengineering; in its 'making' through deliberate process and, in relation to the possibility of democracy. Rather than attempt a systematic examination of the field of geoengineering, the project situates geoengineering in its wider historical and social contexts in order to open questions about how geoengineering might represent a distinctly new type of geopolitical formation, informed by a reconceptualization of geophysical processes rather than distinct geographical territories.

In parallel, this essay examines some of the logics that underpin geoengineering, specifically the reconstituted of the 'geo' as a territorializing force within environments. The contested and speculative nature of geoengineering has generated an expanded set of discourses in both popular and scientific domains. As both a practice and a concept of planetary change, geoengineering brings together new issues to do with world risk, anticipatory governance of futures, atmospheric securitization, innovation and new technologies, and the ethics and politics of "earth systems governmentality." Such an aspirational model of 'whole Earth' governance follows in the wake of a decade of climate governance that has been characterized by global climate modelling, technological and market-led solutions, failure of international agreements on mitigation, and arguably a lack of democratic and devolved decision making.

This reimagination of the possibilities of planetary modification happens within a broader geohistoric moment of the consideration of humans as geologic agents within the Anthropocene. The logic used to defend geoengineering to its critics bucks this enlargement of the geopolitical sphere, suggesting that there is little distinction between inadvertent geoengineering (anthropogenic climate change) and overt climate engineering, just one of intent. However, there is a definite change in the scale and quality of governance and thus the modes of territoriality that are imagined in planetary phase shifts. The shift is from the generation of predictive climate scenarios to predictive interventions in climate actualities. The ambition of earth-system governance to make global interventions in the evolution and revolutions of the earth requires an arguably new formation of the relation between geography and power, which entails both ontological and material shifts in the scope and scale of human agency in biophysical and geophysical earth forces. As such, the possibility of unilateral, yet regionally unequal modification of climate prompts the need to think about a geopolitics that extends its consideration to the bio and lithosphere, in order to think through forms of collaboration within earth processes rather than outside of them.

Like the climate sciences, geoengineering has generated new epistemological and ontological models of thinking about the earth, which simultaneously engender both the imaginaries and operative spaces that calibrate notions of agency and control in earth processes. As political decision making takes on materialist dimensions through the development and testing of these technologies, new geopolitical approaches may be required that depart from classical or critical geopolitics foci on mostly 'flat earth' or vertical political relations to conceptualize the interactions and thresholds of earth systems, biochemical flows, and the forces of deep-time processes, in concert with political decision making. This would require the conceptualization of a geopolitics that considers inhuman forces (anthropogenic and otherwise) alongside more traditional political actors in the making of geopolitical worlds. This is not just the imagination of earth-as-artefact, as in the Apollo images from space or the future-orientated earth objects of climate modelling, but it involves an understanding of the earth as a geoengine that can be altered, modified, and engineered on a global scale. The engineering, then, is not metaphorical, so much as bio-geophysical; it involves the conceptualization of an abstract, nonlinear biodynamic model of the earth in which atmospheres and earth systems, and even the sun, are imagined as 'engines' (albeit organic) open to forms of making. This is not about creating new worlds per se (although geoengineers do now talk about 'ideal climates'), but about a shift from the earth as a given to the earth as characterized by idealized forms of output, such as climate.

While geoengineering is often framed as a mindful response to anthropogenic climate change, the terminology of the Royal Society report, entitled *Geoengineering the Climate: Science, Governance, and Uncertainty* (2009), makes a distinction between what should be considered as geoengineering, and what should not: weather modification and anthropogenic experimentation that have inscribed geophysical 'signatures' into earth (ballistic missiles, combustion engines, atomic testing, acid rain, fire regimes, biomass burnings, industrial revolutions, and deforestation). Part of this 'distinction' is no doubt pragmatic, to set the terms of reference, but it also enacts an erasure of Cold War histories of weather modification—weather as war machine and geopolitical strategy—that generated highly speculative (and nearly always unsuccessful) experiments. This historical amnesia distances geoengineering from the rationalist male fantasies of modernist control that permeated Cold War projects, while reinstating the ambition for global reach or technological transformation that sustained their promise. What geoengineering gains in loosening its historical ties to weather modification projects of the 1950s is an ability to claim its precedence, as a 'new' solution to climate change. What it perhaps fails to recognize in its Cold War genealogy is how the promise of former technological innovations already mark how politics are made in the desirous gap between intention and actuality.

The geologic of innovation (located within a broader field of eco-innovation) is underpinned by a temporal metaphor—geoengineering will buy a little time—for adaptation in the wake of abrupt climate change. Yet, this is no quick fix. And, it is a technological fix that needs to keep on fixing. Modelling of interventions suggests that the rebound effect of not continuing along a geoengineering trajectory once started would be catastrophic. This raises questions about what exactly is the worst-case scenario when geoengineering is unidirectional in its deployment, and commits societies to a continued technology intervention into the far-flung future. Like nuclear waste, geoengineering has a technological shelf life that far exceeds the intent and foresight of its contemporary engineers. So, geoengineering practices may 'live on' beyond their milieu. Such a contingent futurity of a new earth, which may be characterized by abrupt geophysical and social change, raises questions about intergenerational justice and the intensification of risk in the context of uncertain futures. The temporal quality of commitment to geoengineering technologies in some instances is irreversible, and yet this commitment is undertaken under the temporal conditions of a state of emergency to establish the validity of its claims.

In the embrace of the emergency conditions ('considering the unthinkable') there is not just a commitment to the continuance of geoengineered futures—the earth as anthropogenic entity—but to a potentially antidemocratic and unethical passage through climate change. Rather than finding ways to rethink how climate science policy could and should be done differently, geoengineering proponents mobilize this inaction to argue for interventionist policies. When the politics of description involved in defining what geoengineering is gets reimagined in the context of entrepreneurial business solutions to climate change (the main funder of geoengineering technologies to date), the language becomes decidedly militaristic (the "geoengineering battlefield" is part of Richard Branson's "Carbon War Room") and muscular (Shell calls geoengineering a "Game Changer" and the Bill and Melinda Gates Foundation calls it a "Grand

Challenge"). This 'emergency' looks decidedly like the same business-as-usual approach to expand and generate new market economies of carbon management.

Furthermore, geoengineering technologies raise ethical questions about the biopolitical impact of geoengineering on nonhuman life, and demonstrate how these geotechnics are becoming central to the governance of life beyond the human. Currently, international jurisdiction is governed by the United Nations Convention on the Prohibition of Military or Any Other Hostile Use of Environmental Modification Techniques (1997), which states that "the term 'environmental modification techniques' refers to any technique for changing—through the deliberate manipulation of natural processes—the dynamics, composition or structure of the Earth, including its biota, lithosphere, hydrosphere and atmosphere, or of outer space" (Article II). But it further qualifies this by stating, "the provisions of this Convention shall not hinder the use of environmental modification techniques for peaceful purposes" (Article III), which suggests that geoengineering has to be proved to be warlike or geopolitical in intent to counter the Convention. Real-world experimentation also raises questions about what it means to move from lab to world, and from model to actuality, and how the sensibility of an engineered planet enables understandings of human agency that has ambitions to act as an earth force.

Within geoengineering research, disciplinary assumptions have shaped the problem formation and failed to take a properly interdisciplinary approach, and so subordinate the social in addressing what geoengineering could be. This is the imaginary of a machinic earth. While notions of a machinic earth have their genealogical origins in James Hutton's *Theory of the Earth* (1788, page 3), it may be that this metaphorical association that immediately gives rise to the figure of the geoengineer as the agent of control is unhelpful in capturing the diverse set of material practices that characterize the scope of geoengineering's current oeuvre. The work of Design Earth invites us to consider how an interdisciplinary approach might change the questions that are asked, the practices that are imagined, and the modes of governance that are engaged with. What possibilities exist for an engagement with geoengineering, which highlights not just the ethics of geoengineering, but also engages with the processes and practices of problem formation, including its inclusions and exclusions? Could we imagine a more diverse set of governance practices and earth interventions, where publics are not just involved in giving dissent or consent to geoengineering projects, but can interrupt, experiment, and intervene in problem formation? What modes of participative engagement would be necessary to establish a dialogue rather than a monologue with publics?

At present, geoengineering— the intentional modification and/or management of the earth's climate system—remains, "largely speculative and unproven" (IPPC, 2007), with limited real-world experimentation. The praxis of climate making in geoengineering might be characterized by what Mike Fortun (2005) calls in relation to the Human Genome Project, "promissory technologies." These are technologies that are characterized by the promise to be transformative in the future in ways yet to be known, but are often subject to "over-promising" in terms of their potential. In a revisiting of the ethics of promising, Fortun suggests that his initial attempt to critique such promises by biotech companies was limited, because he did not remain open to the promise. He says: "Promising cannot be reduced to either empty hype, or to formal contract, but occupies the uncertain, difficult space in between."[1] Fortun reminds us that, in trying to understand the discourses that emerge with the geoengineering technologies, it is not necessary to proceed from a place of judgment on the technologies themselves, but rather it may be better to find ways of understanding the discursive and geopolitical affects that the promises of those technologies mobilize to transform understanding of what these things are or could be. That is not to abandon critique, but to engage with geoengineering as an object of concern and praxis of understanding through modes of speculation. Such new geoformations might challenge both our conceptualization of geoengineering and help us rethink the material 'ground' of planetary geopolitics.

Geoengineering too needs its promissory and geopolitical intent creatively explored if we are to move from exceptionalism to more participative forms of governance. The arts and humanities may well offer particular critical investigations into these technologies. Artists, designers, writers, and theorists in the Cold War period, working between the specter of nuclear catastrophe and utopian architectures for planetary survival, offered both playful and critical geopolitical accounts of imaginaries of atmospheric control (such as Buckminister Fullers's

1960s “Dome over New York”). Such design begs a model of knowledge making that is both critical and creative in its exploration of the norms and structures of knowledge-making practices in geoengineering research. By engaging publics with the processes, technics, and imaginaries of geoengineering, the arts open up questions about the making of models and worlds. In this sense, cultural experiments in narrating geoengineering involves a form of geontologising: that is, the reconstitution of the earth as a dynamic world object.

While geoengineering is often framed predominantly as a governance issue—a way of managing science in society—the fundamental geophysical nature of interventions into the geoengine are somewhat obscured. What are the geopolitics and speculative geophysics that might be necessary in order to think at the scale of the planet? While geographers might assume a certain priority in thinking the earth the focus on the politics of space, have left the geophysical aspects of earth systems woefully underthought in social worlds. Taking up geoengineering as provocation at the edges of political possibility, and in the speculative spaces of a new geopolitics of earth processes is a way to think a fidelity to the full promise of geoengineering—opening a space for thinking about geologic agency within earth forces—in order to forge a new geopolitics that includes the actual geophysical earth as constitutive of its world view.

This essay is an adapted version of "The Geoengine: Geoengineering and the Geopolitics of Planetary Modification," first published in *Environment and Planning A* 2013, volume 45, pages 2799–2808.

Note
1. Fortun, Mike. 2005,“For an Ethics of Promising, or: A Few Kind Words about James Watson,” *New Genetics and Society* 24(2)157–173, 158.

Geoengineering a Rare Earth (for Rare-Earths)

Benjamin Bratton

Benjamin Bratton is Professor of Visual Arts at the University of California, San Diego and Program Director of The New Normal programme at Strelka Institute of Media, Architecture and Design in Moscow. He is author of *The Stack: On Software and Sovereignty* (MIT Press, 2016) and *The Terraforming* (Strelka Press, 2019).

I wish to question the term “geoengineering” in relation to what it is that is being engineered, by whom, where, when and why. In a way, the notion is not too hubristic but too myopic. My own questions, however, has to to with the position of intelligence from which such intervention might take place, how that position might comprehend the situation of its agency, and, finally, how it comprehends the planet represented by the prefix “geo.” I can’t answer them, but I hope that I can at least ask them properly.

First, I would step back and look up (which is really out). For much of the twentieth century, our solar system and galactic neighborhood was apparently full of alien civilizations with whom we were about to make contact. Or so it would seem based on the hypotheses, presumptions and hopes of science and its popularizers. Carl Sagan often mused that should Earth make such contact, determining conclusively that we are not alone, but rather part of a vastly richer fabric of life than we could have ever imagined in our evolutionary infancy, that this would be the most profound scientific and philosophical event possible. I think it likely would be. The entire figure-ground perspective of reality would bend in a moment. The sky would no longer be just the sky but now something else, something more oceanic. The fetid insularity of our theisms would explode into incredibly more generous embraces of the numinous scope of the real. Cosmology could finally begin, and therefore, in a way, so could humanity.

However, over the course of my lifetime—a half-century blip—a different conclusion has made itself more clear and pressing. Having sent probes past the nearby planets and their moons we find violent gases and dead rocks, and as those probes extend out of the Sun’s grasp they find an empty void. Only since the Berlin Wall came down have humans made the monumental discovery and initial cartography of thousands of exoplanets. Some of these could, in theory, support life, and some of those could, in theory support complex life, and some of those could, in theory, support a civilization, and some of those, in theory, could support a civilization capable of communication, but all of these are so far away that the transit of a message from there-to-here or here-to-there might take longer that humans have been making cities let alone radio telescopes. Most of what we really know about our cosmic situation has been gleaned in the last generation and the bracing conclusion to be drawn aligns with the Rare

Earth hypothesis that life and especially intelligent civilizational life is incredibly unlikely and incredibly rare. That is: we are very, very likely alone. It is not that life doesn't exist elsewhere, but that the modest arc of hominid evolution is not likely to intersect with it, especially not during relatively instantaneous moment we call history.

If the reckoning with contact would be the most significant philosophical event imaginable, is the reckoning with the likely impossibility of contact -that we are functionally alone—also itself equally profound? I think it is. A certain kind of mature cosmology and sense of planetary predicament comes into focus in a way that is more necessary. And yet, the dawning confirmation of this cosmic isolation has not been met with explosions of insight or rapturous fear, but rather with a shrug.

Instead of a shrug, however, what really are the implications for us and what may be our responses? Really, we can't know, at least not yet. It will take the rest of our existence to construct them. Some, though, are immediate and pretty obvious. Firstly, Astrobiology is not the study of extraterrestrial life but the condition of Earthly life as one particular form of terrestriality. What is terrestriality? To be determined. Less obvious questions follow on. Given the coincidence of the evolutionary emergence humans and human technologies, how to make sense of the historical couplings of reason, expression and violence? What are the implications of extinction, not just for humans but for holocene ecosystems? Not only would the only biodiversity within trillions of millions vanish but so would the only form of life capable of knowing that this is the only life within trillions of miles.

In terms of how we conceptualize that prospect, Thomas Moynihan asks us to make the crucial distinction between apocalypse and extinction. Apocalypse is an eschatology for which the narrative arc of history is conceived as the fulfillment of an underlying worldly ethics; it is a Judgment Day when and where the intrinsic ethics of Creation are resolved. It is an end that humans would not only witness but would, in most tellings, play a starring role as hero, villain or victim. But what if the universe is not itself ethical, and there is no arc and no narrative beyond what we impose? Human extinction is the end of the form of intelligent life that is capable of comprehending the very idea of extinction, of planetarity, of the extinction of other species, and of the very premise of ethics. This is much more chilling. Apocalypse is the meaning of the end. Extinction is the end of meaning.

Human intelligence is not just something that happened on Earth; it is something that the Earth is. Over millions of years, this particular planet has folded itself in such a way to form mammalian brains and eventually human sapience, which eventually came to ascertain its uniqueness and its aloneness. It is not only capable of remaking the planet in the image of its industries, both petty and mighty, but also to comprehend and conceptualize the significance of this fact. In that conceptualization, it has come to a fork in a very long road. Is the revelation of its position of agency a cause for abdication of that power, given the destruction that the angel of history retroactively surveys? Or, is this revelation of consequences a kind of threshold of maturation after which, this intelligence—a human intelligence because a planetary intelligence—might now, finally, be able act back upon the planet with care, precisely because it now bears the ponderous weight of its precarious isolation?

There is a tradition, extending from East to West and back again, that has embraced the challenge of the latter possibility. It includes the Cosmist dreams of Konstantin Tsiolkovsky but also the authors of ancient geoglyphs, the architects of space age land art, and the technicians of CRISPR/Cas9 gene editing. These intersecting threads share an understanding that humans and their planet are not separate, that human intelligence and planetary intelligence are not separate, and that humans and the planet are both mutable. The last of these informs a sense of purpose: human/planetary intelligence not only can be a position of agency for how one remakes the other, but that this intelligence appeared and evolved through the exercise of that agency. This agency is not only something that such an intelligence possesses; it is, finally, what that intelligence is.

What we call today "geoengineering" is, at once, the descendent of these heterogeneous threads and also an outlier. It is a realization of some of their ambitions and yet, in present form, also operating as if strangely ignorant of them. To clarify this point, we would need to differentiate a general and specific connotation of the term. The general connotation is geoengineering as a scale of design effect. Anything done (or not done) that has "geoscale" effects could, in principle, qualify. The specific connotation would refer to techniques like solar radiation management (SRM) and direct air carbon capture (DAC). DAC is a negative emission technology (NET) that demands orders of magnitude more interest, investment, attention and leverage than it presently receives. SRM is another story. Unlike, for example, CRISPR, which may open up a deep horizon of new general therapies and elective augmentations, anyone with a serious sense of the conditions into which SRM would intervene hopes that it will never have to be used. At the same time, that serious sense also comes with the sober knowledge of what one degree, two degree or three degree warming really means. Each means scenarios that not only should be avoided but which must be prevented. They are futures that must be prevented. And so, there is no way to explain away the risks of such interventions nor to explain away the risks of not intervening in these ways. There is no comfortable position where one can stand without getting dirty.

The worst such scenarios include the prospect of human extinction and the slightly less worse feature the complete collapse of civilizations, past (as archives), present (as our known reality) and future (as the generational project). These too must be prevented by any means necessary. The Rare Earth hypothesis makes explicit that as those things slide away into the mud, so does the exceedingly rare, difficult, unlikely, precious accomplishment of planetary/human intelligence. The life yet to be lived demands to live. "But it would return," you might say. Some present artifacts might make for good insect architectures, and something else may, in time, perhaps, return to fill the so-called "cognitive niche" on our planet, one that can support the formation of intelligent life but might do so necessarily, like Jupiter supports gaseous storms. But, how long would it take for the planet to generate a form of life, carbon or silicon, that is capable of comprehending that this extinction even happened? How recent and contingent was it that humans could recollect the Permian Great Dying on behalf of what came before?

What has brought us here? To suggest that the general connotation of "geoengineering" is an inescapable aspect of the human vocation, and that a serious reckoning with the potential applications of specific techniques is prudent, is not to also suggest that either the general or the specific are in some way "good." There is also no approximation of the "good" that does not include an encounter with them. Put differently, there is no way to realize the intelligent ethical composition of the world that is not also "geo" in scale and tempo, but there are many ways to apply that intelligence that are, paradoxically, also its own betrayal.

Nothing portends more for the comprehension and construction of a viable or non-viable planetarity than the prospect of planetary-scale computation. It may all stand or fall there. The very concept of "climate change" is among the accomplishments of this accidental megastructure. Its relationship to the long term future project of "geodesign" and of "designing Earth" is then at least two fold. First, it mediates the conduction of carbon and silicon based intelligences, working in some collaborative, centuries-long mutual birthing. This complex allows for contemporary Earth Sciences to describe past and present conditions and to calculate confident simulations of the future; it allows for an awareness of an Earthly condition and the means to act back upon it. Computation is, in other words, both the medium through which our position of agency acts and also itself another position of another kind of agency, both trained, we hope, on a viable planetarity.

However, if understood as itself a folding of planetary materiality into a composed form, planetary scale computation is also an artificially-distributed mineral ecology no less than the the Amazon is a biotic ecology. It lives off minerals and for what? Let us take rare-earths elements (REE) as exemplar. These soft heavy metals are sought after for their magnetic and electronic properties. They are, despite the implication of their name, rather plentiful geologically speaking. China's ban on export of REE has fixed a near monopoly on their value chain, and there are significant environmental issues directly related to its mining and processing. They are fraught in every way imaginable, and our collective computational cognition, at least momentarily, now depends upon them.

Harking back to my original set of questions, it is impossible for me not to observe a dark parallel between the emergence of human sapient intelligence, bound as it was and is to the abstractions of war and politics, and the emergence of computational intelligence, bound as it is to the wounds of extraction and ecological degradation. Both overlapping positions of reason are built of not only aspiration and inspiration but also of gore and stupidity. If planetary intelligence is to succeed, it must heal the conditions of its own appearance. It is born into a crisis that it was born to address.

This moment is defined by the attempt to make sense of the evolutionary predicament of being sapient as a rare and precious planetary circumstance and of how this intelligence is itself seen in the reflection of the violence and destruction that is indissoluble from its own emergence. The question asked by that reflection looking back at us is whether this violence was a necessary condition of the emergence of planetary intelligence, and even if so, if whether the growth and expression of that intelligence for the future is now to account for and graduate from that violence? Perhaps the futurity of planetary intelligence is now just as inextricable from the disappearance of irrational destruction as its evolutionary appearance was from the rolling waves of destructions that made it possible?

I can't say which I believe. I am not entirely convinced that destruction is avoidable, even as I recognize that I will be among what is destroyed for the sake of life's continuance, and for the maturation of planetary self-comprehension and self-compositional agency. When considering the dark scenarios, I think about the Svalbard seed bank. I consider what other such archives should demand from us their creation. What would it take to ensure that if extinction is the only real accomplishment to come, that there will remain a durable, protected, and revivable form of planetary intelligence that would at least know that this extinction happened, and that could comprehend it and conceive of what comes next and what to build instead.

While we still can, we may want to imagine ourselves as occupying the position of agency of this archived survivor intelligence. We may want to presume revision as the starting point. I can't say that this revision would be without destruction, but what is certain—perhaps the only thing certain—is that the violence of extinction that erases the possibility of future life, reason, intelligence and ethical abstraction is, by far, the greater violence to be refused by any means necessary. In my most optimistic moments, I think of "geoengineering" as a name for this refusal.

Finally, the question behind the question implied by the Fermi paradox is not just 'where is everyone?' but rather 'what is a civilization and what are its potential parameters'. Asking how we might recognize an artificial signal from another galaxy becomes instead 'what is artificiality, what must artificiality be on and for this planet, this sapience and this technical reason?' These are questions, among others, that respond to the "philosophical event" of the revelation of the Rare Earth. They are questions without definite answer; they are fundamental and permanent. Fermi is forever.

Grounding Climate Engineering (In Five Stories)

Holly Jean Buck

Holly Jean Buck is assistant professor in the Department of Environment & Sustainability at the University at Buffalo in Buffalo, New York. She is author of *After Geoengineering: Climate Tragedy, Repair, and Restoration* (Verso, 2019).

The first time I saw geoengineering visualized by an artist was in Biosphere II, the old algae-streaked glass edifice in the Arizona desert. In the early 1990s, eight scientists came together to try to live for two years inside a complex dome, as if in Spaceship Earth: ocean, desert, rainforests in miniature. The biosphere's ecosystems fell out of balance, with insufficient oxygen and species die-off. Inside the biosphere remains a brief warning about geoengineering; a panel with six black-and-white tiles representing different geoengineering methods. Three of the tiles present antiquated drawings: a steamer ship for ocean iron fertilization, a naturalist's rendering of seaweed, a billowing cloud for "Burn thousands of tons of sulfur to stimulate formation of clouds that would screen sunlight." The other drawings are geometric forms: space mirrors, reflective balloons. Black boxes: but stylized to be as if from the past, we *don't dream of grand designs anymore.* The title was "Techno-fixes", as if the run-down biosphere was trying to tell the visitor something. Biosphere is the old hand at the bar, nursing its drink, *don't make the mistakes that I did.*

The stories of "The Planet After Geoengineering" also act as warnings: not from the past, but from the future as it is unfolding. To be clear, climate change is not a story; 415 parts per million CO_2 in the atmosphere is not a story. But saying "we need to cut emissions from fifty billion tons a year to zero by midcentury", or saying "we need to remove between a hundred billion and a trillion tons of carbon dioxide from the atmosphere by 2100," is too vast to comprehend. It is impossible to wrap one's mind around the material, social and political upheavals without a story. Curbing warming to 1.5°C will require quickly slashing greenhouse gas emissions by building vast new energy infrastructures, changing how food is produced, planting new forests, redesigning buildings, changing how we move around.

It is possible that we succeed at this work, but unlikely— is it impolite to acknowledge this, or is it dishonest not to?—and because of this, scientists, tech philanthropists, and groups of climate-alarmed civilians are looking into the possibilities of "enhancing earth's natural heat shields", or taking "actions to re-instate natural climate processes" or "buying humanity more time by actively cooling the planet." In other words, solar radiation management or geoengineering, stripped of the stories and context that Design Earth sketches here.

The late atmospheric chemist Paul Crutzen set the tone for this in a 2006 essay in the journal *Climatic Change*, positing geoengineering by stratospheric aerosol injection as a possible response given how unsuccessful climate action had been to date, and opening the idea to wider scientific debate. Today's new hopeful grift doesn't say the word geoengineering, because the word is too terrible, for reasons that are palpable in these stories. *Geoengineering* brings with it every dark part of the twentieth century and beyond: the military violence, the gridding of the world and erasure of the non-gridded, the accumulation and extinction. No wonder people want to write a new story for it, a "climate restoration" story or an emergency medicine story. Those stories, while attractive in that spunky action-hero way, risk losing sight of the social and spatial forms of power that helped get us into this predicament.

Key questions around geoengineering usually come in a certain order. There are the pragmatic ones: Can we intervene? How? What is most effective? What is cheapest? If attention isn't disrupted, then there are moral questions: Should we intervene? Then, question of power: But who decides? Who designs? Visual and multimedia forms can lead us into these questions, disrupting the sequence of them, creating space around the technologies-as-artifacts or infrastructures.

Petrified Carbon is rooted not both in the subsurface and in the Industrial Revolution. This is climate intervention according to the logics of dark satanic mills, of production taken to the extreme: the whole biosphere is put in service of maintaining fossil fuel production. Carbon capture and storage doesn't have to be "a path to further dependence on the fossil fuel industry," as the narrator warns—this is but one implementation, one serious story. It is one we should listen to, given the actors who are developing this technology and their legacy.

Sky River offers another resonance for the story—geostory as myth—with ancient magical practices, gods and goddesses. The ritual rain dances present geoengineering as a kind of magic, which it may turn out to be. It will be very hard to "know" how geoengineering is working, because of the complexity of earth systems; it is a challenge to parcel out the effects of one intervention. In that sense, geoengineering is in practice a ritual act of faith.

Story as caution, story as myth: but there's a third type of story that creeps in, which is that of conspiracy, making a theory about how the world works and storifying it. The High-frequency Active Auroral Research Program (HAARP), referenced in *Arctic Albedo,* studies the ionosphere and has been long suspected by conspiracy theorists of being a weather control experiment. They too imagine its eerie hum, relentless in the polar night. This is the dark terrain of chemtrails, of 5G arrays, of microchipped vaccines, of primal evil organized that wants to destroy life and control minds. People in this community produce art of their own: wastelands of antennae and carcasses, the sky carved up with chemical lines, children with syringes in them left by the wayside. The modern myths made on the internet aren't simply dismissible, though; they have something to say about the ecological crisis. While the platforms are developing algorithms to inoculate the public against these stories—newly alarmed by the potential for terror that believing in these stories might lead to—repressing the stories will lead to disaster, too.

That's not to say that everything is terrifying. Stories also bring ambiguity, though: *Arctic Albedo* also brings the lab-grown mammoths managing the boreal forest, which are also sublime. When the constant hum of the scrubbers in *Petrified Carbon* transforms the city into a machine, an industrial plant—with resonances of the nineteenth century as much as the twenty-first or twenty-second—is that purely terrifying? Or a success? The story, particularly the embodied experience suggested by the sounds of the burbling permafrost or the pings of industrial exploration, forces us to consider these speculations apart from the heuristics of rejection or acceptance. The stories engage our curiosity about what living there would actually feel like; push us into deeper consideration of how to salvage the inspiring parts and parcel them out from the terrors.

Sulfur Storm introduces another element of the story: the character. We meet several faceless men around a table, Men Who Are Deciding; we glimpse the actors that have sent these machinations into motion, and wonder about the psychology of these characters. Edward Teller and his nuclear ambitions are referenced as one lineage for the psychology. This story, more than any other, highlights where the psychology leads: to the dome as division, where 1.5 billion people live inside the privileges of the geoengineering, inside the vivarium, with billions more outcasts outside. While at other points the encapsulated worlds in the stories are almost enchanting—snow globes, underground cities—here the dome presents its stark divisions made material. The ultimate wall separates those who can live within a habitable climate from those without. Not so much "the planet" after geoengineering, but society after geoengineering, and in *Sulfur Storm*, the situation is traced directly back to the characters: someone was responsible for this. To put something into story form is to afford these assignations; to make the causal connections.

Finally, these are also war stories—"wars on life," as *Dust Cloud* points to. If we want to talk about a war on climate change, if there is truly a "climate emergency", these are the kind of stories we get: air raid sirens and the rest of it. As intoned here: "The terror of geoengineering is to be forced to live intimately with the death of the atmosphere, contemplating suffocation as a real possibility." The action is on the battlefield; the battlefield is everywhere, outside us, in space, inside our bodies.

These are the stories of intervention in the climate within the 20th century imperialist, capitalist logics leads. Geoengineering cannot be divorced from its context, say these stories. Current discourse around geoengineering has taken the technologies and practices of climate intervention out of the world, wrapped them up in special paper, and demarcated them as novel medicine. Saving the earth is going to require a deep transformation of how we use energy, how we manage waste, how we construct shelter, how we grow forests, how we produce food, and more. It is by thinking of these things separately that we are in this predicament. Intervening in the climate is not some separate topic; it is a relatively small part of this everything-transformation.

Fiction is a possible method to counter the realist conventions of geoengineering: speculative technologies demand speculative narratives. The geostories here present the folly of parceling out parts of the system; they demand we address the embodied lives and emotions of people who live in the future, as well as the groups that get to bear the risks from particular implementations. They counter the violent ways in which spatial imaginaries of geoengineering void human and nonhuman life.

Some may resist stories like these; they may say, these stories are unfair to the possibilities of using geo-intervention technologies for collective flourishing. Or: *That's not what I mean by modeling futures, and it's not my fault the models use grid cells.* To them I might say: the only way out of antiquated ways of thinking about climate control for fossil capital and the military-industrial complex is to understand how they have shaped our imaginations and to trouble visions of what is possible and desirable. Once we do that work, then perhaps we can invent other stories about what comes after geoengineering, of worlds worth living.

The Planet After Geoengineering

Project Team
Rania Ghosn + El Hadi Jazairy, Reid Fellenbaum, Monica Hutton, Kelly Koh, Meng-Fu Kuo, Anhong Li, Joude Mabsout, Jane Jia Weng; with Ayusha Ariana, Avery Nguyen, Michael Stradley.

Project Sponsors
Massachusetts Institute of Technology Center for Art, Science & Technology
University of Michigan Taubman College of Architecture and Urban Planning

Commissioner
Biennale Architettura 2021, titled "How Will We Live Together?" curated by Hashim Sarkis

Published by Actar Publishers, New York, Barcelona
www.actar.com

ISBN 9781948765961

Library of Congress Control Number: 2021935565

Distribution: Actar D, Inc.

New York
440 Park Avenue South, 17th Floor
New York, NY 10016, USA
T +1 2129662207
salesnewyork@actar-d.com

Barcelona
Roca i Batlle 2-4
08023 Barcelona, Spain
T +34 933 282 183
eurosales@actar-d.com

Design assistance, Office of Luke Bulman

Printed by Hampden Press, United States